A Buyer's Guide
To Olive Oil

Anne Dolamore

GRUB STREET • LONDON

This book is for all those people who care passionately about the quality of food,
who want to learn about really good olive oil and are prepared never to settle for second best
for the sake of price.

ACKNOWLEDGEMENTS

I wish to thank all the importers, producers and distributors who generously supplied me with
samples of oil to taste and bottles to photograph for this book, as well as much useful information.
Special thanks to Mary Contini of Valvona and Crolla, Mark Lewis of Harvey Nichols, Pat Harrison
of The New Cook's Emporium, Ledbury and Nick Reitmeier of Selfridges for taking time to supply
me with their stock lists. Maria Jose Sevilla Taylor of Foods From Spain for organizing such useful
trips to Spain, Luis Avides of ICEP and Edite Vieira for arranging the trip to Portugal, Daphne
Derven of the American Center for Wine, Food and the Arts for inviting me over to Napa and
Rosemary Barron for guiding me round while there. The International Olive Oil Council in Madrid,
for information on the latest medical research. Giles Henschel of Olives Et Al for information on
PDOs and PGIs. Panos Manuelides of Odysea, Charles Carey of The Oil Merchant, Petra Carter in
Ireland, Clarissa Hyman, Catherine Brown in Scotland, Gilli Davies (and her book *Eat Well in
Wales*) for recommendations of good shops in their areas. Dominic and Naomi for being great
assistants and putting up with an office full of boxes of oil. Simon Smith for photographing the
cover and Alan Batham for photographing the bottles. Julie Adam for coming up with the excellent
cover design and last but not in any way least, John and Amy for giving me the space and time to
write and taste, and for making all the toil worthwhile.

Published by Grub Street
The Basement, 10 Chivalry Road, London SW11 1HT

British Library Cataloguing in Publication Data
Dolamore, Anne
A buyer's guide to olive oil.- New ed.
1.Olive oil
I.Title
641.3'463

ISBN 1 902304 23 3

Printed and bound in Spain by Bookprint, S.L., Barcelona

CONTENTS

FOREWORD

It is six years since I compiled the first edition of *A Buyer's Guide to Olive Oil* and I am delighted that the book proved such a success and became the bible for shoppers and shop owners alike, supporting my feeling that there was a need for guidance through the bewildering choice of oils on the market. The purpose of the book was confirmed every time I walked into a food shop and saw a well-thumbed, oil-stained copy open in front of the oil display for customers to consult. Robert Cockcroft, kindly wrote, when reviewing the first edition in *The Yorkshire Post*, that the guide might do for olive oil what CAMRA did for beer by my outspoken comments on oils, both favourable and unfavourable, and it was most satisfying to see how my views were taken account of when importers of some of my favourite oils used my recommendation in their advertising.

For the last few years since copies of the old edition became unavailable I have been constantly asked by people to write another edition, but pressure of work as a full-time publisher prevented what is a marathon undertaking of organizing samples and tasting day after day. However over the intervening years I have run tastings and generally kept up to speed on developments, written many articles and given numerous radio interviews on the subject. I knew the time had come to work on a new edition when I spotted a woman in a supermarket whose shelves were groaning with bottles of extra virgin olive oil from Sicily, Tuscany, Crete, Liguria, Umbria and Mani; she was reading with furrowed brow three bottles in particular. The bottles stated they contained in one Picual, in the next Hojiblanca and the final one Arbequina. I knew she didn't have a clue what this meant. She told me she thought they were places. These are the two major changes that have occurred since the last edition, the recent proliferation in the shops of regional extra virgin oils and the marketing of single varietal oils, such as Arbequina, Picual, Hojiblanca, Taggiasca, Frantoio and Coratina. So there is even more reason than ever before to offer some guidance on what to find in these bottles.

I have sniffed and tasted over 150 different extra virgin oils for this book. Having done over 100 previously for the old edition there were oils I knew I liked and looked forward to tasting again, some were as good as in the past but some were no longer as good as they were and of course there were masses of new oils never tasted before, always an exciting experience. Some general and useful observations have emerged in the course of all this tasting as my knowledge on the subject continues to grow. There is always something new to learn every day from all this research. One real illuminating flash came earlier this year when I was invited by the Consumer's Association to be a judge for a blind tasting of seventeen extra virgin oils for an article in WHICH? Magazine. This was a truly depressing experience because apart from one or two oils, all of the samples tasted were old, tired and flat. They were oils in very poor condition. Why should this be? I suddenly realised that when I taste oils for an article or my books I have samples delivered to me direct from the distributor. The oils tasted for WHICH? had however been purchased straight from the shops. They had been transported round the country and had been sitting on a shelf for who knows how long, maybe in full sun, maybe in a window, certainly under bright lights. So it's not enough for me to taste oil and say whether it's well made and delicious, you the consumer may not come to taste it in its best shape. The fact is unless you know a shop turns their oils over quickly you may well be buying, at not inconsiderable cost, a bottle of tired old oil. Certainly what also became apparent from the WHICH? Tasting was that some of the oils were a blend of last year's and this year's oil. This is a practise I deplore. It simply gives you a poor, flat tasting oil instead of vibrant, fresh tasty oil. Alas I think this is something that is likely to become more prevalent in the future with bulk-blended oils. This is because there is so much extra virgin oil around now and many producers who formerly would have sold out of their oil at the end of a year have lots of oil still in storage at year-end, when the new harvest commences.

I have also come to realize that the other factor, which plays a major role in the condition of the oil you buy, is the 'best before date'. Olive oils have a two-year best before date. This is quite simply far too long a time period from the pressing date. Extra Virgin is at its best about six months after pressing. Depending on the olive variety it should be still OK but mellower after a year from pressing. So forget any oil if you see it has a 'best before' of six months or less to run. Best of all look for those enlightened producers who print the pressing date on the bottle.

Shopping for any commodity, if you care about the integrity and quality of what you consume, is sadly becoming more of a minefield. Though on the one hand it appears we have more choice than ever before, the debasement and corner cutting in the world of mass-produced food can be a problem for the discerning shopper. This is especially true of olive oil. Just because it says first cold pressed Extra Virgin olive oil made from hand picked olives from a single estate on the bottle doesn't guarantee, I'm sorry to say, a wonderful product. This is certainly true if it is supermarket own label.

However all is not doom and gloom as I hope you will find on reading through my tasting notes. There are some truly exceptional oils available and it has never been easier to obtain the really good oils because of mail order and numerous web sites offering dazzling selections of produce. I must stress though that of course taste is very subjective and you may not always agree with my opinions. I have to confess to a preference for more assertive, characterful oils but even where I don't like the style of an oil I will acknowledge if it is well-made. The scourge of my wrath is reserved for poorly made, flabby, greasy old oil. I hope my book will just help to guide you to decide for yourself what you like, whilst helping to eliminate the bad and indifferent, because what you like is what matters. As I always say when asked about what is the best oil, it's the oil *you* like best.

INTRODUCTION

Pliny the Elder's sound advice that 'a happy life is one that uses wine inside and olive oil outside' will certainly ring true today with those of us who have discovered albeit two thousand years later, the pleasures and health benefits of the Mediterranean diet. Though we may limit our use of olive oil solely to the culinary rather than anointing our bodies, lighting our lamps or greasing our chariots. Ancient history and literature is filled with references to olives and olive oil but if it's the history and romance of olive oil you are looking for then I must refer you to my previous book, *The Essential Olive Oil Companion*. This relates the ancient and fascinating story of the olive; when it first appeared and how its history is woven into the myths and legends, as well as the everyday life of the Mediterranean. It also includes over 100 recipes incorporating olive oil, collected from friends and acquaintances in the olive growing countries over many years.

The aim of *this* book is to provide a practical and informative guide. I shall explain exactly why olive oil is now medically proven to be good for you, what olive oil is, how it is made, the different types, different styles, what affects the taste, and what information is to be found on the label so that when you go to buy olive oil you will know how to make informed choices and having purchased your oil how to use it to best advantage.

I always have at least three different grades of olive oil in my kitchen. There will be several bottles of the very best single estate or single varietal extra virgin for its fabulous variety and intensity of flavour. These I use to drizzle raw over cooked dishes such as soups, casseroles, pasta, pizza, cooked vegetables, bruschetta, salads, salad leaves, and finally for lazy lunches or barbecues with friends as a dipping oil on the table for bread. I like to have as large a selection as possible of estate oils because I use them according to my mood, the occasion or the dish, just as I might splash on a different perfume every day. Next, I will always have a bottle or a 3 litre can of commercial blend extra virgin at the cheaper end of the price range for frying, basting, making sauces and everyday cooking, and finally some olive oil (this is the name, rather confusingly, for refined olive oil, and used to be called pure olive oil). This olive oil even though it is refined is very useful and can be used whenever you may cook with seed oil, such as sunflower oil or peanut oil. People still have this misconception that olive oil is no good for deep-frying and revert back to other oils. I only have olive oil in my kitchen, I think there is an olive oil for every kind of cooking and every dish. I know some people disagree with me saying they prefer other oils for neutral flavour, but there are plenty of light, mild olive oils for these occasions and because olive oil is a natural food full of minerals and vitamins, I prefer to use it rather than the aggressively refined seed oils, which have to be treated to remove their natural toxicity.

So that's my store cupboard of olive oils and I would suggest if you want a basic 'start-up kit', you arm yourself with the same: an estate or single varietal extra virgin, a commercially blended extra virgin and olive oil. But you may not be sure what these terms mean or why it is better for you to use olive oil in your diet.

OLIVE OIL AND HEALTH

I find a basic knowledge of the chemical composition of olive oil makes it easier to understand the concepts relating to the present high esteem olive oil holds medically.

Composition of olive oil
Vitamin E (3-30mg)
Provitamin A (carotene)
Monunsaturated fatty acids (oleic) 56-83%
Polyunsaturated fatty acids (linoleic) 3.5-20%
Polyunsaturated fatty acids (linolenic) 0-1.5%
Saturated fatty acids 8-23.5%

9 calories per gram

The body can produce most of the fatty acids found in foods but it cannot make either linoleic or linolenic acid. These are termed essential fatty acids and as you can see from the table above olive oil contains both of these making it an important and valuable dietary ingredient.

Olive oil contains a preponderance of oleic acid, which is a monounsaturate. Fatty acid proportions vary in olive oil depending on region, olive variety, year, harvesting etc. Besides being a monounsaturated fat, virgin olive oil, because it is a completely natural untreated food, is rich in antioxidants and vitamins, which help prevent body cell ageing, as well as giving the oil itself conservation properties. The chlorophyll content, which gives a green colour and the carotene content, which gives a reddish pigment on ripening, lend each oil its final colour. The volatile aromatic components influence odour and flavour. Polyphenols account in part for flavour but due to their antioxidant nature they have a decisive effect on the stability of oil, in other words on its keeping properties and resistance to degenerative tendencies.

Olive oil and heart disease
Oil and fats are essential in a well balanced diet, as they are a rich source of energy. But which type of fat we consume has been shown to have a direct bearing on the incidence of a range of diseases now found with increasing frequency in the developed Western world. In general those people living in the Mediterranean eating a typical Mediterranean diet live long and healthy lives.

They are less likely to be obese, they have fewer heart attacks and lower incidence of breast cancer, bowel cancer and diabetes. Medical research supports the assertion that their health is due in large part to their diet. Mediterranean diets are high in cereals, vegetables, legumes, fruit and fish, with small quantities of meat and dairy products. Olive oil is their major source of fat and the foods they eat are mostly eaten fresh rather than being processed.

In stark contrast the typical Western diet is low in cereals and vegetables, legumes, fruit, nuts, red wine and olive oil; foods which provide high levels of antioxidants. It is a diet high in sugar and saturated fats found in fast food, processed food, meat and dairy products.

OIL AND FAT COMPOSITIONS

	% Saturated	% Monounsaturated	% Polyunsaturated
Coconut oil	92	6	2
Olive oil	12	80	8
Corn oil	16	27	57
Sunflower oil	10	18	72
Safflower oil	12	10	78
Butter	58	39	3
Margarine	64	30	6

All this goes back to some of the most remarkable nutritional research studies carried out in the 1950s, which showed that in Greece and Crete, with their low intake of saturated fat there was the lowest incidence of heart disease and that though the people living in these regions had high total fat intake, the fat came from only one source – olive oil. As a result people in the West were urged to consume less saturated fat and thus began the huge campaign promoting polyunsaturated fats. Even though the studies had shown olive oil to be beneficial the polyunsaturated bandwagon had begun to roll. Until finally in the 80s research showed that monounsaturated fatty acids (in which olive oil is especially high) are more desirable because they selectively lower cholesterol levels. This startling research came up with amazing evidence about the nature of cholesterol.

There are two types, low density (LDL) and high density (HDL). Low density lipoproteins (LDL) transport and deposit cholesterol in the tissues and arteries. LDL increases with excessive consumption of saturated fatty acids and is therefore potentially harmful because it will deposit more cholesterol. HDL on the other hand eliminates cholesterol from the cells and carries it to the liver where it is passed out through the bile ducts. Polyunsaturates reduce both LDL and HDL but monounsaturates reduce LDL while increasing HDL. An increase in HDL will not only provide protection against cholesterol deposits but will actually reduce cholesterol levels in the body.

Attention has focused recently on antioxidants since medical research has shown that cholesterol in the blood is damaging when it oxidizes. Antioxidants prevent this occurring. Olive oil contains a great range of different antioxidants and extra virgin oil has somewhere in the region of 30 to 40 different antioxidants. Some other oils contain the same quantity of antioxidants as olive oil but not the range of different protective antioxidants. So when LDLs oxidize they form many oxidation products and these can only be neutralised by a range of antioxidants.

Olive oil and breast cancer

Olive oil's virtues however may go further than just protection against coronary heart disease. Some of the polyphenol antioxidants in olive oil may have the ability to destroy substances that lead to the proliferation of cells in the development of cancer. Further research is needed to work out how this protection occurs but evidence shows that women in Mediterranean countries have much less breast cancer than the high rates in such countries as the United States and Australia.

Olive oil and infancy

Since olive oil also contains Vitamin E and provides a relatively low amount of essential fatty acids but has a balanced linoleic and linolenic ratio similar to that found in breast milk, it aids normal bone growth and is particularly suitable for expectant and nursing mothers because it encourages the development of the infant's brain and nervous system before and after birth. Seed oils, which are rich in polyunsaturates, are not recommended in large quantities for children because it is not advisable to lower their cholesterol level.

Olive oil and ageing

The defence mechanism that Vitamin E provides through its antioxidant properties is also vital in the ageing process. Mice fed on olive oil have a longer life expectancy than those fed on sunflower and corn oil. This is explained by the better ratio between Vitamin E and polyunsaturates in olive oil. Bone calcification is another problem common in the elderly. Olive oil seems to have a positive effect, which appears to be dose-dependent, because the more olive oil ingested the better the bone mineralization obtained. According to French researchers, olive oil would appear to be necessary during growth and later in adulthood to avoid calcium loss.

Olive oil and diabetes

Non-insulin dependent diabetes is increasing at an alarming rate in Western countries. There is some evidence that it might be related to the kind of fats within cell membranes and the way the membranes resist the action of insulin. An excess of the type of polyunsaturated fats found in many vegetable oils and most margarine may be one of the causes of the problem. Olive oil on the other hand provides a better balance of fats within the cell membranes. The other reason olive oil is recommended for diabetics is of course the advice given to sufferers to consume less saturated fat and as we have seen olive oil is low in saturated fat and high in monounsaturates.

From all this evidence doctors and nutritionists recognize virgin olive oil as having the most balanced composition of fatty acids of all edible vegetable oils. It is also becoming increasingly clear that a return to a Mediterranean style diet, high in dietary fibre, low in saturated fats and high in antioxidants (of the type which olive oil affords) is linked to optimum health.

OIL VARIETIES

There are so many different varieties of olives, varying according to the country or even region and like grapes they have different qualities and flavours. Some of the more common ones are given in the table below.

COUNTRY	VARIETY	USE
Argentina	Arbequina	Oil
France	Aglandeau	Oil
	Cailletier	Oil
	Grossane	Oil
	Picholine	Oil
	Tanche	Oil
Greece	Koroneiki	Oil
	Kalamata	Table/Oil
Italy	Coratina	Oil
	Frantoio	Oil
	Leccino	Oil
	Moraiolo	Oil
	Nocellara del Belice	Oil
Portugal	Cobrançosa	Oil
	Galega	Oil
	Madural	Oil
	Verdial	Oil
Spain	Arbequina	Oil
	Cornicabra	Oil
	Hojiblanca	Oil
	Lechin	Oil
	Picual	Oil
	Picudo	Oil
	Verdial	Oil
	Manzanilla	Table
United States	Mission	Oil/Table
	Manzanilla	Table

So the olive variety will provide differences of taste and fragrance in your bottle of olive oil but what else? Olives will vary in taste and colour from year to year, and even from one pressing to another. There is no such thing as a vintage in the wine sense with olive oils but of course some years will be better than others because of the weather. All of the following have influence:

- variety	-cultivation	-method of harvesting
-soil	-health of the fruit	-extraction method
-climate	-degree of ripeness	-country of origin

The beauty of the olive tree is that it can survive where other crops cannot grow. So many of the olive groves planted in the past were on land that was poor quality, located on hillsides where erosion had washed away the most fertile layer of soil or in areas that received very little rainfall. The fact that olive trees can grow under such inhospitable conditions does not mean they do not respond favourably in a more fertile, better-irrigated environment. It is true that while little has changed over thousands of years in producing olive oil, modern technology is providing the olive grower with the means of improving his productivity. Foliar analysis during the winter rest period and soil analysis are now regularly used to determine any deficiencies, which can in turn be rectified by fertilizers or applications of organic matter. Soil management, pruning, micro-irrigation and crop health controls are all vital to the modern olive producer and will determine the health of his olives and hence the quality of his oil.

The timing of and method of harvesting and how the olives are handled after harvesting will have, in my experience, the greatest impact on the resulting olive oil, more so than the method of extracting. If the olives you take to the mill are over-ripe, infested with olive fly, damaged by frost, bruised from falling to the ground and stored for days in piles of sacks no matter what you do the oil will be bad.

Despite countless efforts to develop mechanized harvesting methods I still maintain that the best olive oils are made from hand picked olives at the optimum level of ripeness which usually means while they are still green and starting to change colour, and certainly before they start falling from the trees.

If olives are shaken from a tree and fall to the ground they bruise and bruised fruit will start to oxidise and ferment, raising the acidity level of the resulting oil and as you will discover later on, acidity levels are the criteria by which olive oils are categorised. Mechanical harvesting because it depends on shaking the fruit from the trees inevitably causes the ripest fruit to fall, leaving the greener fruits still clinging on. So you harvest consistently blacker fruit rather than the ideal mix you get with hand gathered fruit. Each variety has its peak of maturity, at which stage the olives need to be harvested in order to extract the best oil. Generally the best oils are made from olives harvested when they are between one third and two thirds black. Olives which are too green give an intensely bitter taste and those which are too ripe or black produce oil with a flat taste and often a high acidity.

If you've ever wondered at the price of the finest extra virgin olive oils then the fact that they are hand picked should reveal all. Imagine the labour involved, and thereby the cost, in going over an entire olive tree, and not just one olive tree but thousands upon thousands of olive trees and you'll have some conception of why you'll pay dearly for the best olive oil, but it will certainly be worth every single pound. And by the way it takes 5 kilos of olives to produce 1 litre of oil and each tree can produce between 3 to 4 litres of oil. Just think if you're prepared to spend £6 for a bottle of very good wine, maybe even £20 for an excellent bottle which will last you just one meal or one evening, then consider that a bottle of olive oil at a comparable price will last you months. Unlike wine however it will not improve with age, so don't bother laying it down – it should really be used within the year.

Harvest time, depending on the olives and the country of origin, is between September and December but may go on to January or even February. The olives are finally taken to the mill and the quicker they are processed the better. The best olive oils are crushed and pressed within a few hours of picking and no longer than 24 hours. However if the weather is exceptionally cold they may be stored for a while to allow them to heat up a little, helping finally to release the oil from the crushed fruit, but certainly for no longer than a day or so, otherwise they will start to ferment, which would ruin the resulting oil.

The olives are then washed to remove leaves, twigs or earth, and crushed to produce a homogenous mixture from which the liquid can be extracted. The traditional method involves grinding the olives with a granite wheel. In recent years a hammer mill has replaced this. There are advantages to each method. The stone wheel is slower and more gentle but more difficult to keep clean. The hammer mill on the other hand is cleaner but tends to pulverize the olives, imparting, some people think, a rather metallic taste.

The paste obtained by crushing the olives is kneaded mechanically to help the amalgamation of the minute droplets of oil found in the pulp. This resulting mixture is a combination of liquid (oil and water) and solids (pulp and stones). Between 96 and 98 per cent of the oil is contained in the flesh, with only 2 to 4 per cent in the stone. Roughly 15 to 25 per cent of the fruit is oil, whereas the quantity of water varies between 30 to 60 per cent. Of the rest 19 per cent is sugar, 5 per cent fibre, 1 per cent protein and the rest various mineral elements. There are two basic methods of extraction. The first is called traditional and involves the extraction of the oil by mechanical pressure. The pulp of the olives is spread in thin layers, separated by round filters made of synthetic material. These are then placed in a hydraulic press with pressure varying between 250 and 400 kg per square centimetre. The oil runs off from the solid matter, filters through and drips into containers, where the oil and water are separated most often these days by means of a centrifuge. As with the granite stones using mats raises the problem of cleaning to prevent rancidity. They must be steam cleaned regularly.

While the traditional press can produce extremely fine oils the need for constant cleaning has caused it to fall out of favour with large-scale producers, who tend to prefer the continuous-process centrifuge, or decanter. Centrifuges are efficient and clean and virtually all new, modern facilities use centrifuge. Here the paste is spun at high speed to separate the mixture of oil, water and suspended solids. Both methods are equally good, though many people think continuous extraction gives a greater consistency of quality; others consider the method too violent causing the oil to emulsify.

Another method of extraction, which has been much in vogue in Italy, is the Sinolea machine. After the olives are crushed the olive paste goes into the Sinolea, which has thousands of vibrating stainless steel blades; there is no pressure this is all done by vibration, the oil floats to the surface of the paste and is drawn off. Some people claim this produces the very best oil, I have yet to be convinced, but it only releases about 30% of the oil produced by pressing or centrifuge so it is an expensive method. The theory behind the claims for its greatness as a system is it produces next to no heat and is so gentle.

When the oil has been separated from the water you are left with first cold pressed virgin olive oil, which is a totally pure product because it is untreated. Cold pressed means that the temperature during the oil extraction process has been controlled not to exceed 30°C. If the temperature rises above this level the

quality of the oil will be affected. No other vegetable oil is edible just by being pressed. All other oils have to be treated first because they contain toxins or are not suitable for human consumption in their natural state. Once the oil has been separated it is usually left to marry for a while, then bottled. If the scale of production is large enough, the oil gets bottled a little at a time because it keeps better in bulk than in bottle. True state of the art installations store the oil in stainless steel tanks filled with inert gas to occupy the space left when oil is drawn off. This prevents any contact between oil and air, which will encourage rancidity.

GRADINGS OF OLIVE OIL

So what is the oil in those holding tanks at the producer's mill? And how does it compare to what you eventually buy? When you come to choose a bottle of olive oil you will see labels bearing the words *extra virgin*, occasionally *virgin* and then *olive oil*. To really understand what each of these are it helps to know that virgin olive oil is graded according to its acidity in the following categories set down by the International Olive Oil Council.

VIRGIN OLIVE OIL is the oil obtained from the fruit of the olive tree solely by mechanical or other physical means under conditions and particularly thermal conditions that do not lead to alterations in the oil. Further the oil has not undergone any treatment other than washing, centrifugation and filtration (i.e. not refined). Until 1996 extra virgin olive oil simply meant mechanically extracted oil with an oleic acidity of less than 1%, Virgin Olive oil was mechanically extracted with an acidity of between 1% and 3%. In 1996 the IOOC added a new requirement for extra virgin status, of perfect taste and aroma. Oils are put through a blind organoleptic analysis by a panel of highly trained experts.

Given the importance of classifying the quality of virgin olive oils for consumption it seems worth outlining exactly what goes on. A panel is made up of between 8-12 expert tasters and headed by a supervisor. The supervisor marks each sample with a code so that the tasters do not know what oils they taste. 15ml of oil is placed in a blue tasting glass. This is so they will not be influenced by the colour of the oil. The temperature of the oil is about 28°C, this is deemed to be the ideal temperature to detect all aromas and flavours. The Method lays down the criteria for recognition of the oil's attributes. Their positive and negative intensities should be indicated (positive – fruity, bitter and pungent flavours; negative – fusty, musty, muddy, winey/vinegary, acid/bitter, metallic, rancid etc). The Method also sets out a common vocabulary so that all tasting panels will uniformly understand these terms. Each taster sits in a booth, alone in a quite room. He completes a Profile Sheet with his olfactory, gustatory, tactile and kinaesthetic (pressure applied to the sample in the mouth) perceptions of each sample, noting different qualities and defects. He also evaluates the intensity of each of the positive and negative attributes using a scale of 0 (minimum intensity) to 10 (maximum intensity). Each sample has its own Profile Sheet and the supervisor calibrates a matrix of the results.

VIRGIN OLIVE OIL fit for consumption as it is, is classified into the following:

EXTRA VIRGIN OLIVE OIL is virgin olive oil of absolutely perfect taste and aroma (when the defects average from the panel is 0 and the average fruitiness is more than 0) and a maximum acidity in terms of oleic acid of 1%, though many of the best are 0.5% acidity.

VIRGIN OLIVE OIL is virgin olive oil of absolutely perfect taste and aroma (when the defects average is between 0 and 2.5 and average fruitiness is more than 0) having maximum acidity in terms of oleic acid of 1.5% or less.

ORDINARY VIRGIN OLIVE OIL is virgin olive oil of good taste and aroma (when the defects average is between 2.5 and 6 and the average fruitiness is 0) having a maximum acidity in terms of oleic acid of 3%.

VIRGIN OLIVE OIL not fit for consumption as it is, is classified into the following :

VIRGIN OLIVE OIL LAMPANTE (lamp oil) is an off tasting and or off smelling virgin olive oil with an acidity in terms of oleic acid of more than 3.3%. It is intended for refining and becomes one of the following:

REFINED OLIVE OIL is olive oil obtained from virgin olive oils by refining methods.

OLIVE OIL is a blend of refined olive oil and one of the top three grades of virgin olive oil to give flavour and aroma. It has acidity in terms of oleic acid of no more than 1.5%. This oil under the old regulations used to be called Pure Olive.

So you can see from the above that the lower the acidity level the better the quality of the oil. The acidity of oil is important also because it affects the speed at which the oil will deteriorate. The only necessary qualification however for a virgin olive oil to rate being called extra virgin is its acidity and here is the rub, as it is possible to reduce acidity by a variety of methods, which means lesser quality oils can be promoted into what should be the premier league. Some extra virgin olive oils may not be all that they should be.

It is a commonly held myth that you can tell what an oil is like by its colour. For the record colour, which can range from yellow to intense green gives absolutely no indication of taste or quality, which is dependent on olive variety and degree of ripeness. A green olive will give a different coloured oil to a ripe black olive and while I'm on olives you may also not know that there is no such thing as a green olive variety and a black olive variety as with black and white grapes; every olive starts green and when it's ripe it's black. So forget about colour because in some places they crush a few olive leaves with the olives to make the oil greener! The only hint you may glean about an oil from its appearance is that if the oil is opaque or hazy, there is a pretty good chance it hasn't been filtered, which is fine as it simply means there are bits of olive in the oil.

My message therefore, when buying extra virgin olive oil is *caveat emptor*.

BUYING OLIVE OIL

Extra Virgin
Estate grown and bottled extra virgin or single olive variety extra virgins from hand-picked olives are the best olive oils you can buy and as a consequence they are at the top end of the price range. They can cost anything from £10 – £30 per litre. The reason for the high price is the quality you are getting and the great care that has gone into producing them, often in very limited quantities. They are not produced in bulk and are usually made by the people who have grown the olives and had them pressed in their own mill. They are the equivalent of top growth wine or finest champagne, and mostly I would advise it's best not to cook with them, (unless of course you are lucky enough to be able to afford to). I'm fortunate to always have a ready supply of the finest extra virgins and I do confess to cooking with them when flavour really counts. However I do think they are best used as a condiment, as a heavenly flavouring poured over salads or dishes after they are cooked, so that the heat of the dish brings out the gorgeous flavours. I almost never mix them with vinegar or lemon juice, because to my mind their flavours are so exquisite on their own they need nothing more. After all you wouldn't think of mixing premier cru claret with lemonade would you? If you are amazed or appalled by the idea of using finest olive oil on just its own then you haven't yet explored this essential ingredient. The most sublime of these expensive oils usually come from Italian estates. Some of my

favourites come from Umbria, Liguria, Tuscany and Sicily and the varietals to look out for which are my favourites are Taggiasca, Nocellara and Coratina. There are also a few equal quality oils from Spain especially from the DO area of Siurana in the north, made from Arbequina olives and Baena in the south.

Commercially Blended Extra Virgin

An important distinction to be aware of with extra virgin olive oils is what we term commercially blended extra virgin. These extra virgins can be likened to *vin de table*. They are the brands you will see in almost every supermarket and independent grocer such as Berio, Sasso, Napolina, Bertolli, Cypressa and Carbonell, as well as, of course, the supermarket's own-brand extra virgin oils. These companies buy in olive oil by the tanker load from local producers and co-operatives and blend them. They will, as a consequence, taste much the same year in year out because, just like non-vintage champagne, they are blended to a particular style. These oils are considerably cheaper than first cold pressed oils because they are produced in bulk, the average price being £6 per litre.

These commercial blends of extra virgin are everyday oils. Use them to make salad dressings with lemon or vinegar or other flavourings. They are great for marinades with herbs and spices or basting food while it is cooking, such as chicken or grilled fish or general cooking where you might otherwise use a seed or vegetable oil. Some of them are of good quality and thereby exceptional value for money, but sadly most of the supermarkets seem to stock their shelves with the same ubiquitous brands whose quality often leaves much to be desired. At the moment some of the best value-for-money extra virgin olive oils in this category come from Spain and Greece.

One other twist to this complex question of precisely what are you getting in your bottle of extra virgin oil, is the present EU legislation which allows countries to import olive oil, bottle it and re-export it without having to declare the country of origin of the oil. Italy imports Spanish and Greek olive oils in enormous quantities, blends them with Italian olive oil and sells it abroad as produce of Italy. This is not a criticism and there is nothing underhand in this practise, it's just that you need to realise that just because it says Italian on the label it may not be that the oil has actually originated in the country. This is one reason now to start looking out for oils that indicate region of origin. These are essentially the old Denominations of Origin or DOs. Under new legislation these are now Protected Designation of Origin (PDO) or Protected Geographical Indication (PGI).

Here is a list of the current registered PDOs and PGIs, which may help you when selecting oils to buy:

FRANCE
(PDO) Huile d'olive de Nyons

GREECE
(PDO) Archanes Iraklion Crete; Apokoranas Chania Crete; Kranidi Argolidas; Krokees Lakonia; Lygourio Asklipiou; Petrina Lakonia; Peza Iraklion Crete; Sitia Lasithi Crete; Viannos Iraklion Crete; Vorios Mylopotamos Rethymno Crete
(PGI) Chania Crete; Hania Crete; Kalamata; Kefallonia; Kolymvari Hanion Kritis; Lakonia; Lesbos; Preveza; Olympia; Rhodes; Samos; Thassos; Zakynthos

ITALY
(PDO) Aprutino Pescarese; Brisighella; Bruzio; Canino; Cilento; Collina di Brindisa; Colline Salemitane; Colline Teatine; Dauno; Garda; Laghi Lombardi; Mont Iblei; Penisola Sorrentina; Rivera Ligure; Sabina; Terra di Bari; Terra Otranto; Umbria; Valli Trapanesi
(PGI) Toscano

PORTUGAL
(PDO) Azeite da Beira Interior; Aziete da Beira Alta; Aziete da Beira Baixa; Aziete de Moura; Aziete de Tras-os-Montes; Aziete do Norte Alentejano; Aziete do Ribatejo

SPAIN
(PDO) Baena; Les Garrigues; Sierra de Segura; Siurana

The Italians have gone one step further and have introduced a High Standard or HS. The association of Maestri Olivari and the Oil Masters Corporation started the initiative. The objective of this new category is to safeguard producers of extra virgin olive oil by introducing production parameters even more severe than those applied by the European Community. HS oils besides the characteristics stipulated for extra virgin olive oils must possess superior nutritional characteristics, a high profile sensorial level and an assured minimum antioxidant level. The HS quality trademark is granted by an international body and can be applied for by single producers. It is felt that this trademark ought to give clearer indications to the consumer about price quality levels rather than competition based on price only.

Supermarket Oils
In my experience of tasting all the extra virgin olive oils for this book there is rather too much indifferent quality commercially blended extra virgin oil on the market, especially among the supermarket brands, with none of the fresh aroma and taste which should be found there. The problem with the supermarkets is

they buy solely on price. They have seen the olive oil sector of the market grow and the demand from the public continues unabated. Years ago when I first started writing about olive oil, supermarkets stocked one or two brands. As it became a fashionable ingredient, they increased their range. Then they got into having their own brands; at first just any old Extra Virgin, then they realised people were starting to appreciate the national differences in the flavour of olive oil, so they introduced Spanish, Italian and Greek. Now they've moved on again into the regions; Tuscan, Cretan, Sicilian and of course if they can stick Organic on the bottle so much the better. Considering about three years ago the supermarkets reckoned there was no demand for organic produce, they've certainly lashed their carts to the organic gravy train in a big way. Only they would have us believe it's because they are listening to their customers and providing choice. Don't be fooled, they are interested in high volume sales and profit, quality is a secondary factor to price when it comes to supermarket fare and the olive oils they offer are no different. Apart from a couple of notable exceptions, from Marks and Spencer and an excellent organic oil from Asda, which I have written about in my tasting notes, I wouldn't out of choice buy supermarket brands. The tired, flat, or tasteless liquid they pass off as extra virgin olive oil is an insult. And it's obvious why. They have about three price points for extra virgin oils; £2.99, £3.99 and £4.99 for 500ml. In order to fill these price slots they find the cheapest oil around. In most cases what they buy I can only assume the producers were only too happy to see the back of. The supermarkets dump on us inferior quality extra virgin oils that would never be sold in Italy, France or Spain. The Germans buy most of the best extra virgin oils because they are prepared to pay for quality. One insider told me that there is oil in Spain, already two years old, which is being purchased for some of the supermarket own brands. If that oil then has a two year 'best by date' it doesn't take much maths to work out that you could by the end of that period be using oil over four years old. The other thing that annoys me about their oils is the way they dress them up in fancy bottles and then describe them in flowery prose to have us believe in some cases that they are estate bottled oils, with lots of waffle about hand-picked olives, limited quantity, etc. No wonder the public are confused when they read on the label that the oil is made from first cold pressed finest olives and is rich and fruity, or full of flavour when the oil is actually a flavourless liquid with about as much taste as tap water.

In any event olive oils from supermarkets are unlikely to be in peak condition. The oils get transported endless miles in the back of warm lorries and could have then stood outside in the sun at a depot waiting to be stored. Once in the shop the bottles will sit under banks of fluorescent lights. Guaranteed to take its toll on olive oil, which needs a certain amount of care in handling and storage.

Where to buy

Go to a shop where there is someone you can discuss your taste in oil with. Plenty of enlightened deli owners these days allow you to taste oil before you buy. Many are now selling extra virgin oils loose so you can buy a small bottle and use it quickly and will always have the freshest oil. If you don't have a good shop nearby there is now plenty of opportunity to buy mail order or over the net and I have given details of all these outlets and rated them in the directory section at the back of the book.

And I do strongly recommend trying to taste whenever you can because oils change from year to year just like wine. Some years are good; some are bad, because of climate, insect infestation, and the weather at the time of the harvest. I have found sometimes that oils I generally like can be disappointing some years. These changes do not apply to commercially blended oils because they aim to blend to the same flavour each year. I am talking here about single estate or single varietal oils. So remember when reading my notes that the harvest you may be tasting could be a different one to the one I have tasted.

COUNTRY BY COUNTRY GUIDE

Every country around the shores of the Mediterranean produces olive oil, as well as Australia, New Zealand, Argentina, South Africa and California and the style of the oils is different in each one. But here I shall just concentrate on the countries whose oils most often find their way into our shops. Generally speaking the oils of France are delicate and subtle, sweet and gentle; in Greece they are grassy and green; in Italy they range from the strongly assertive, spicy and peppery in Tuscany up in the north to the fruitier ones of Apulia right down in the south; in Portugal rustic and earthy and in Spain bursting with the flavours of tropical fruits in the south or nuts and almonds in the north. Though there are within each of the countries strong regional differences also.

France

You may be surprised to discover that France is actually one of the world's smallest producers of olive oil. Consequently, while more French oil is now finding its way into the shops it is not widely available but it has always been held in high regard. The provinces in the south and especially those bordering the Mediterranean make up the twelve departments where olives grow with the best areas considered to be in Provence. Throughout the region you will find hundreds of small olive mills serving the thousands of tiny producers. Nyons is the olive town of Provence, nestling in the foothills of the Alps on the banks of the River Aygues, where the local variety, Tanche, thrives in the microclimate here. The oils are sweet and fruity and Huile d'olive Nyons has its own Appellation d'Origen (now a PDO).

Travelling south towards the coast you come to Maussane beneath the hills of Les Alpilles. This has a famous co-operative, thought by many to produce the best olive oil in the whole of France, with a strong, distinctive flavour sold under the name of La Vallée des Baux and made predominantly from the Picholine variety. One other olive oil of international repute is from Monsieur Alziari and sold from his shop in Nice. Made from olives grown locally it has a sweet and delicate aroma and a flavour to match. Generally if you come across oil marked Huile de Provence you're pretty well assured of quality. The best range of French oils available outside of France is from the Oliviers & Co Shops. They have delicious sweet oils from Haute Provence, Pays d'Aix and the Vallée des Baux, made from Aglandau and Grossane olives. I don't always like the very fruity rustic style of some of the French oils as you will see from individual tasting notes but the good ones are so different and distinctive and deliciously sweet that I strongly recommend you try them if you've only ever sampled Italian oils.

Greece

Greece is the third largest producer of olive oil in the world and the Greeks consume more olive oil per head than any other country. For many years the Greek oils did not find favour with our palate; they were described as rustic and assertive, though personally I always liked them. However the oils coming into the country now have been blended to a lighter, less aggressive style and clearly the Greeks have worked hard at establishing real quality. The main producing areas are the Peloponnese and Crete, where the most important olive variety for oil making is Koroneiki. If you're thinking I've forgotten Kalamata, this is the town in the Peloponnese that gives its name to the best variety for eating. The Greek oils available over here are generally good quality, everyday oils for cooking and salad dressings and are good value for money especially if purchased in 3 or 5 litre cans.

Italy

Without question some of the most sublime single estate bottled oils come from Italy and not surprisingly since Italy has more varieties of olives than anywhere else in the world. Every region produces olive oil, though the world at large may still mostly think of Tuscany and its celebrated town of Lucca. Many of the Tuscan oils are indeed gorgeous, reminding me often in aroma and taste of fresh, green meadows. Up front and assertive with greater or lesser degrees of peppery after taste, sometimes the pepper is too much for me and I find myself coughing from their powerful kick. Oils from this region are inevitably the products in the main of Chianti growers and you will find the gallo nero logo reproduced sometimes on their bottles of oil. My advice with Italy is be adventurous, branch out and try just for a change the oils I have recommended from some of the other regions such as Umbria, which grows many of the same olive varieties found in Tuscany but which are fruitier and less green and bitter; Liguria, whose oils are light and fruity, and Apulia, the country's largest producer mostly from Coratina olives. The oils are delicious and full of the flavour of ripe olives. I have tasted a large number of extra virgin oils recently from Sicily and have been so impressed by the high quality and styles of the oils; fortunately more of them are coming on to the market and you will find that some of my present favourites come from Sicily. Look out for single varietals from Italy too, especially Frantoio, the variety common in central Italy particularly Tuscany, Marche and Umbria. Frantoio is a late ripening olive so olives are greener when picked giving spicy, green oils. Moraiolo, originally only grown in Tuscany is now also cultivated in Umbria. This is an early ripening variety, so the olives tend to be blacker when picked, resulting in softer, fruitier oils. Leccino, originally a Tuscan variety now grown in Umbria, Marche and Latium, often blended with Pendolino and Frantoio; and Taggiasca, widespread in Liguria. But buy any Taggiasca oils as young as you can because it is an olive with the most volatile polyphenols and being a delicate flavoured oil to start with the flavour soon flattens out after six months.

Portugal

When I wrote the last edition of this guide I had only tasted a couple of Portuguese oils and was not in the least taken by them. The style of oils in Portugal is very rustic, flat and earthy, which is because the Portuguese harvest very late when the olives are all black and near to dropping from the tree. However I visited Portugal this year and was impressed by the revolution that is gradually taking place in many establishments. Certain producers have realised that outside of Portugal the flavour of their oil does not find favour. The olive varieties in Portugal are capable of producing fine oils, the common ones are Madural, Verdial, Cobrançosa and in the south Galega but it is traditionally the late harvesting and subsequent long storage of picked olives which have caused the rather off-tasting oils.

So they are harvesting much earlier and pressing the olives much sooner after harvesting. As a consequence there are oils appearing with a fresh, grassy taste very much in the style of Tuscan oils. One of the most delicious oils I tasted in the Tras-os-Montes region will not even be available in Portugal since the very happy, content looking producer informed me that each year a German company buys his entire production. The Madeira family estates produce a delicious range of prize-winning oils and mill under the name Quinta das Marvalhas and they would certainly seem to typify the way things are changing. Up until three years ago the family sent their olives to the local co-operative to be pressed. But they realized their oil could be better. So now instead of harvesting from mid December they are harvesting from the end of October and are endeavouring to keep olives no longer than 24 hours before pressing. They have built their own mill and have the latest state of the art equipment, including a Sinolea machine. As a result they won the national award for best oil last year.

Spain

How gorgeous are the best olive oils from Spain. Oils from the north often have a lovely bitter almond taste while in the south they are packed full of luscious tropical fruit scents and flavours. What is remarkable about Spanish oils is that despite being the world's largest producer there seems to be no sacrificing quality. In fact the Spanish were the first to establish demarcated regions of production, known as Denominacion de Origen, for their olive oils thereby guaranteeing quality. There were four DO areas now PDO areas: the first two are Borjas Blancas and Siurana in Catalonia, where the Arbequina olive tends to prevail. The Arbequina gives a very smooth, elegant oil with hints of almond and green apple. The other two areas are Sierra de Segura and Baena in Andalucia, where the oils are more commonly made with Picual and Hojiblanca olives. Picual olives give sumptuous, intense oil with plenty of body but which is slightly bitter, very distinctive in style. Hojiblanca gives a soft fruity oil. However it has to be said that the quality of olive oil from outside the DO areas is always remarkably good. Most of the olives grown throughout the country are taken to local mills and sold on to huge concerns like Borges and Carbonell. But you will find estate-produced oils such as the famous Nuñez De Prado. You will also find superb PDO oils and I have to say that probably the best extra virgin I tasted this year came from a co-operative in Siurana called Unio made from Arbequina. I could have sat and drunk it by the glass full. It had heaps of flavour and character and is delicious, perfect oil.

United States

Small in world production terms it may be but the investment pouring into creating boutique olive oil ranches in the US is huge, especially in California's Napa Valley. Olives have been grown in California for more than two hundred years. The Spanish missionaries took them originally hence the name of the indigenous variety the Mission, later to be followed by the Manzanilla and Sevillano. These varieties supplied the canning industry with green olives for martinis and black olives (in fact chemically treated green olives) for the pizza industry. But with cheaper imports from Morocco and Spain flooding the market the growers had to think again and so started producing oil. However since the 80s, in much the same way as many of the wineries developed financed by wealthy businessmen who made their fortunes as lawyers or brokers, and wanted to play at being hobby farmers, the Napa is now seeing a rush of small independent growers, who are far from being amateurs, in fact many of these new olive oil producers are the owners of wineries. They take their oil production just as seriously as they applied themselves to learning the wine business. Many of them have been off to Tuscany to learn their craft and there must be more Sinolea machines on the West Coast than in all of Italy. Everywhere you look in Napa you see new groves being planted mostly with Italian varieties. This is serious stuff. To the extent that there is now a California Olive Oil Council, which was founded in 1992 and has nearly 300 members. It has a panel of tasters trained by European experts who are certifying oils. This year the number carrying the council's seal of approval will top 50. The only problem with these oils is the price, as they carry tags of up to $50 a bottle, which is very steep compared to the price of European oils. However they are novelties and are only likely to serve a local populace who like the idea of consuming oil produced down the road. If you are on the West Coast in the autumn look out for and visit if you can one of the many Olive Festivals that have now become a popular part of local life; like the Olive Oils of Americas competition which forms part of the Los Angeles County Fair.

Among the pioneering producers are people like Nan McEvoy of Marin near Petaluma who has more than 12,000 Italian trees growing on 100 acres. I visited this spectacular spread and it's like finding a little piece of Tuscany dropped down in California. They have their own traditional stone mill as well as decanters and centrifuge and they are even propagating their own trees now in an industrial-sized complex of glasshouses. The other estate I visited was Long Meadow Ranch, where Ted Hall the owner produces oil called Prato Lungo. I tasted this first as part of a blind tasting and would have been willing to wager a huge sum that I was tasting the finest example of Tuscan oil. It was thick and vivid green and suffused the nostrils with the headiest aroma and the taste was perfect. What a surprise to discover it came not more than five miles away in Napa. And yet the olive variety grown is not even an Italian cultivar, it is probably related to the French Picholine and the trees were planted back in the 1870s. This is undoubtedly the best Californian oil I have tasted. A number made from Italian varieties I find personally far too buttery. However time will tell which varieties work best in the New World of olive oil production, there are those who feel that Greek or Spanish olives might suit the micro-climate better. Certainly Prato Lungo proves to my mind the success of a French related variety.

Australia and New Zealand

When I wrote my first book on olive oil in 1987, I stated – without fear of contradiction – that olives were only grown around the Mediterranean. Since then much has changed. As I've said above the Americans are fast developing an industry and so too similar events are taking place in Australia and New Zealand. The most incredible statistic is that in Australia, one million olive trees are being planted a year. Just imagine what affect that will have on world supply in five to ten years when the trees are mature enough to produce fruit. Most of those trees come from Olives Australia in southeast Queensland which was set up in 1974. Today it operates what is believed to be the world's largest olive-tree nursery. The mother plot has 28,000 trees which supplies more than 2000 groves across Australia. They grow Manzanilla, Kalamata, Frantoio, Correggiola, soon they will be planting Spanish Picual and Arbequina. Many of the Australian oils made from European varieties lack complexity, but the best flavoured oils, according to grower Maggie Beer, come from wild, South Australian olives: 'Verdale gives a wonderfully flavoured oil but in very small quantities.' This is a feral species that has run wild from plantings by the first colonials in the 1860s, who settled and planted trees which would survive the harsh conditions and provide welcome shade. These trees have been cross-breeding for up to a century and could well be a genetic treasure trove. There is a debate raging in the olive growing fraternity in Australia, and some people feel that it is foolish to be importing European varieties before they have fully explored the possibilities of their own varieties.

It's a similar story in New Zealand. In the 1980s there were only about two producers, now the New Zealand Olive Association has 600 members. The best oil I have tasted so far is a boutique oil from the Albany Olive Press, produced by a passionate grower, Jack Hobbs. I hope it may soon be available here.

Elsewhere

So there is as much happening all round the world at the moment on the olive oil front. They are planting in Argentina, much on the scale as Australia and it is believed there may be as many as 20,000 trees growing in China. Meanwhile the traditional sites of olive growing are increasing their yields by improved irrigation and constantly seeking to improve the quality of their oils. Five years ago only 20-30% of world olive oil production reached Extra Virgin status, now that is a staggering 70%. There is in fact now a shortage of olive oil for refining. Some of the oil for refining comes from Tunisia – the fourth largest producer in the world after Spain, Italy and Greece – as well as Morocco and Algeria. Turkey too produces substantial quantities of oil. But the olive oil from these countries of North Africa and the Middle East mostly supplies the Arab world and little is found in our markets because the style is not favoured by our palates, being strong, rustic and vegetal. Who knows with global warming we may one day be harvesting from the olive trees of Cornwall.

FLAVOUR AND HOW TO TASTE

So what about the wonderful flavours I have mentioned? There are flavours in these extra virgin oils you might never imagine could come from simple crushed olives – a taste of tropical fruit in some, such as melon, lychee or even banana, in others the deep tones of caramelised nuts, chocolate even, green grass, tomato, or apples.

To decide which olive oils suit your palate you need to try a number. So how do you go about this? Well, you can go to olive oil tastings, which are now regularly run by many of the better food shops and wine merchants. Or of course just go out and buy a small bottle, take it home and try it yourself. You will see too from the list of shops, the places that give you an opportunity to taste before you buy.

The way to taste is simple. Just pour a little oil, say a few tablespoons, in a glass. Warm the glass, and thereby the olive oil, in your hands for a few moments to release the volatile aromas in the oil. Bring the glass to your nose and slowly inhale two or three times, taking in the fragrance. It may be complex or it may be simple but above all it should be fresh.

There are then two ways of tasting the oil. You can either dip a piece of plain white bread into the oil and taste it or my preferred method is to take a small sip of the oil from the glass, allow the oil to slide onto your tongue but do not swallow yet. After a few seconds form your tongue into a spoon shape and position it towards your top teeth, now with your mouth slightly open inhale two or three times in quick succession. The mixture of air and oil will spray your mouth and palate, allowing you to register the sensations of the flavours. Try to store in your mind the immediate impressions of sweetness or bitterness, merits or defects. It is then worth repeating the exercise having cleansed your mouth with fizzy water or a slice of apple, the acidity of which will cut through the oil.

There are four main taste experiences: sweet, acid, and salty, which come from the top and sides of the tongue, and bitterness, which is registered by the throat. As the whole palate registers different sensations, it is necessary to swallow the oil to complete your final impression of it which may be different to the first because of the after taste.

At any organised olive oil tasting there should be no more than six oils and this is how I tasted the oils for this book, in groups of six. It is advisable to set out each oil with some in a white saucer to examine the colour and also in a wineglass to enable you to sniff the aroma. You should also supply teaspoons for tasting from and abundant quantities of cubed white bread for dipping, sliced apples, paper napkins for wiping the teaspoons and fizzy water.

Having sniffed and tasted, how then do you describe what you have experienced? There is an official vocabulary drawn up by the International Olive Oil Council for their professional graders and tasters, whose job it is to assess the organoleptic qualities of olive oils and some of these words, which the ordinary taster can also utilise, I give below along with other terms I have coined through my own experience of numerous tastings.

Styles of olive oils vary from the sweet, through the ripely fruity to the assertively green and bitter, with degrees of pepperyness. But remember bitter is not a critical term with olive oil.

BASIC GLOSSARY OF TASTING TERMS

Styles of oils

Aggressive, Assertive or Pungent: oils which have strong up-front flavours or aromas

Bitter: Characteristic taste of oils obtained from green olives or olives turning colour. It can be more or less pleasant depending on its intensity

Delicate or Gentle: a light combination of flavour and fragrance

Fresh: a sensation of freshly squeezed fruit with a significant aroma

Fruity: reminiscent of both the flavour and aroma of sound fresh fruit picked at its optimum stage of ripeness. All sound olives seem fruity after pressing but in most cases this characteristic disappears after a few months. Authentically fruity oil maintains this characteristic through time

Green: often used instead of bitter

Harmonious or Balanced: where the fragrance and taste are in perfect equilibrium. It is the best quality for extra virgin oil to demonstrate

Rustic or Earthy: can be unpleasant or hearty and vegetal

Sharp: a typical flavour of freshly pressed olives, accompanied usually by a brilliant green colour, which fades over the months

Spicy: slight spiciness present in oils in the first months after crushing. Indication of healthy fruit

Strong: intensity of aroma or taste

Sweet: taste or aroma is gentle and graceful, not exactly sugary but found in oil where the bitter, astringent and pungent attributes do not predominate. It gives an initially light sensation accompanied by an almond aftertaste

Aromas and Flavours

Fruity
Apple, Banana, Lychee, Melon, Pear, Ripe Olive, Tomato

Verdant
Artichoke, Bitter Leaves, Eucalyptus, Flowery, Grass, Green Leaves, Hay, Herby, Leafy, Mint, Sorrel

Vegetal
Avocado, Earthy, Rustic

Nutty
Almond, Brazil Nut, Walnut

Chocolatey

Defects

Oxidized/Rancid: the flavour common to all oils and fats that have undergone a process of auto-oxidation caused by prolonged contact with the air. This is an unpleasant taste and cannot be corrected

Soapy, Fatty, Greasy: leaving an unpleasant tactile sensation in the mouth and on swallowing. Often the result of olives that have been attacked by olive fly grubs

Earthy: characteristic flavour of oil obtained from olives that have been collected with earth or mud on them and not washed. This flavour may sometimes be accompanied by a musty-humid odour

Flat: olive oil whose organoleptic characteristics are very weak owing to the loss of their aromatic components, bland and oily

Tired: evident in oils that are too old

OLIVE OIL AND ITS USES IN THE KITCHEN

Having bought your olive oil consume it well before the 'Use by' date on the container. As I mentioned earlier and I can't stress this too much, the finest extra virgin oils should be used within a year after pressing, to capture their maximum flavour and character. Store it away from light and heat because they cause oxidation. You can keep it in the fridge but I don't advise this because it will start to solidify at low temperatures. And likewise if you see bottles of oil in a shop with a white solid layer at the bottom, don't worry – there is nothing wrong with it, this will only be because it has just been taken from a cold store.

If you do buy an oil which doesn't have much flavour then don't despair – remember you can always turn it to good use by putting some herbs or spices in the bottle, such as a few sprigs of fresh rosemary, some peeled cloves of garlic or a small handful of dried chillies. You can experiment with whatever you have available but if you do use fresh herbs take them out of the oil after a week or they'll go mouldy because they aren't sterilized.

You can fry, sauté, stir fry and deep fry with olive oil. Many people say they never deep fry with it, feeling that somehow corn oil or sunflower oil are better but in fact the smoking point of olive oil is about 210°C, which is higher than corn oil (160°C) or sunflower oil (170°C), so you can use a commercially blended neutral tasting extra virgin or refined olive oil. You can filter olive oil after frying and use it again, because olive oil is stable at high temperatures and so unless you have burnt the oil, in which case you should always throw it away, it can be used a number of times. As I mentioned before but I think it's worth repeating, don't cook with your very best extra virgin oils, use them raw, as a flavouring on uncooked and cooked dishes. The beauty of pouring olive oil over cooked dishes is that the heat of the dish warms the oil thus releasing the aroma and flavour to best effect.

So when a recipe calls for olive oil what should you use? As a general rule cook with olive oil, virgin oil or cheaper extra virgin and drizzle with the best extra virgin when the food is cooked. Light or delicate dishes such as poached or grilled fish, chicken and mild flavoured soups use a mild, less fruity oil. Robust, gutsy dishes such as hearty casseroles, stews, tomato and meat sauces, steamed vegetables, salads and bitter salad leaves can take your fruitiest, most characterful oil. For roast, barbecued and braised dishes, use mild or fruity depending on what flavour you want as the end result.

You can use olive oil for baking; as it has a small fat crystal it gives a fine textured, moist crumb in pastry and cakes. You can substitute it for butter or margarine in carrot or fruit cakes, corn bread, batter, pancakes, blinis, crepes and pizza dough. Don't forget too that a slug of olive oil in the water when you are cooking pasta will stop it from sticking together. And of course the ultimate instant snack is to toast bread, rub a clove of garlic on one side and maybe a rub of tomato if you want, then drizzle on some of your favourite olive oil. Delicious and healthy.

France:

Huile D'Olive

Greece:

Mani

Italy:

U Trappitu Intenso

Portugal:

Conservas Rainha Santa

Spain:

Unio

USA:

Prato Lungo

Supermarket:

St Michael Tuscan

Value for Money:

Asda Organic

National Brand:

Carbonell 1866 Organic

Price range code:

Ⓨ less than £10 per L   Ⓨ Ⓨ less than £20 per L   Ⓨ Ⓨ Ⓨ more than £20 per L

A L'OLIVIER

Appearance......Clear and bright

Colour......Golden

Aroma......Heady, rich ripe tropical fruit

Taste......Light and sweet with a lingering flavour of olive

Uses......Cooking, salad dressings, soups, lentils

Price range......Ψ Ψ

COMMENTS
Easy, tasty everyday French oil.
(OM)

FRANCE

LE VIEUX MOULIN

Appearance......Clear and bright

Colour......Gold

Aroma......Cheesy and meaty

Taste......Rustic, and tasting rather of saucisson sec

Uses......Anything with lots of garlic

Price range......Ψ Ψ

COMMENTS
This is prize-winning oil made from Tanche olives grown in Provence. I know lots of people love this oil and a few people have said to me that I ought to like it but I'm sorry I just can't get on with this style French oil. I prefer the sweeter more delicate French oils like Alziari, rather than this 'meaty' style. If your taste is for the rustic style this is for you.
(OM)

ALZIARI

Appearance ......Clear and bright

Colour......Yellow gold

Aroma......Deeply fruity with a back hint of apple

Taste......Bright, light, sweet and almondy

Uses......Raw on salads, meats, fish, vegetables and for cooking

Price range...... ⚱ ⚱

COMMENTS
Fresh bright oil – it's cheerful if that doesn't seem a strange description. The kind of oil that you can use everyday for any number of dishes because of its easy style. Excellent value.
(OM)

HUILE D'OLIVE

Appearance
Clear and bright

Colour
Golden

Aroma
Almost honeyed, sweet and fruity
with a hint of nuts

Taste
Sweet and fruity with a finish
of bitter greens

Uses
Pour it over salads, meat, vegetables
or cook with it

Price range
Ⓥ

COMMENTS
This is such a fabulous example of Southern
French oil under the Fresh Olive Company of
Provence label. It is rich, delicious, fruity and
characterful, with balance and harmony. The
taste makes it very versatile. Good value too
for an oil of this quality.
(FOC)

FRANCE

AEGEAN HERITAGE

Appearance......Clear and bright

Colour......Green gold

Aroma......Lovely immediate green apple

Taste......Smooth, light, very mild but fruity with an attractive lingering taste of bitter leaves. A strong peppery finish

Uses......General cooking and salad dressings

Price range......

COMMENTS

A very pleasant, mild green tasting oil made from hand-picked olives grown in the hills of western Crete.

CYPRESSA

Appearance......Clear and bright

Colour......Green gold

Aroma......Immediate green grass and olive fruit

Taste......Smooth light, very mild flavour with a tang of fresh grass but not at all bitter

Uses......Ideal for cooking, marinades, salad dressings

Price range......

COMMENTS

Very light oil, pleasant example of an everyday oil for all cooking uses. They also produce a very good value Kalamata and Organic oil from the Peloponnese.

GREECE

GREEK ARTISANS FOR SAINSBURY'S SPECIAL SELECTION

Appearance......Clear and bright

Colour......Golden

Aroma......Fresh, immediate aroma of cut grass

Taste......Creamy, smooth, a little pepper on the finish but no real discernible flavour

Uses......Cooking and salad dressings

Price range......

COMMENTS
The label states this is early harvest oil from the Chalkidike in Northern Greece. Why then, oh why, is the colour gold and not green as it would be if it was early harvested oil, or is it just old and has lost its colour and flavour? Not recommended.

ILIADA ORGANIC

Appearance......Clear and bright

Colour......Yellow gold

Aroma......Very faint hint of green leaf

Taste......Very delicate and subtle flavour of green grass

Uses......Cooking, salad dressings, grilled fish and meat, vegetables, salad leaves

Price range......

COMMENTS
This used to be my favourite Greek oil. Its style was very green and grassy; everything that typified for me what was unique about Greek oils. The style is now much lighter and while it is still excellent quality oil, I'm sad about its new subtlety. I preferred it more robust. It's a good value everyday organic oil.
(ODY) (OM)

ELANTHY

Appearance
Clear and bright

Colour
Green gold

Aroma
Lovely intense fresh green grass
and green apple

Taste
Smooth, creamy light, mild green
bitter oil

Uses
Cooking, salad dressings and marinades

Price range

COMMENTS
Useful commercially blended oil made by the
biggest olive producer in Greece. Also
available in 3 litre tins, which is excellent value.
Usually to be found sold at many of the county
and agricultural shows round the country,
which is handy if you don't have a good olive
oil shop near you. See page 126 for mail order.

KARYATIS

Appearance......Clear and bright

Colour......Dark gold

Aroma......Mild hint of green leaves

Taste......Lightly green and grassy, tickle of pepper on the tongue and throat

Uses......Cooking, salad dressings, marinades, grilled meat and fish

Price range......

COMMENTS
Very pleasant light Greek oil. Versatile with just enough character and flavour to make it an easy everyday oil.
(ODY)

OLIVES ET AL KALAMATA

Appearance......Clear and bright

Colour......Green gold

Aroma......Hint of green leaves

Taste......A tickle of pepper on the finish, very mildly green but can't really catch on to any flavour at all

Uses......Salad dressings, grilled fish and meat

Price range......

COMMENTS
Greek oils in my opinion should be obviously green and grassy, this is a disappointing and rather untypical example.
(OEA)

GREECE

MANI

Appearance
Clear and bright

Colour
Green gold

Aroma
Sweet and fruity

Taste
Lovely smooth, light, sweet fruity oil

Uses
Pour over salads, salad leaves, cooked vegetables, grilled fish, bean dishes

Price range

COMMENTS
Made from certified organic Koroneiki olives harvested in Mani, the southern Peloponnese, by traditional millstone and press. It is a delightful fresh fruity oil to use raw.

A MANO

Appearance
Clear and bright

Colour
Golden

Aroma
Delicate aroma of rocket leaves
with a peppery tingle

Taste
Soft and smooth. Green and bitter.

Uses
Raw on salads, soups, vegetables,
pasta and bruschetta

Price range

COMMENTS
Oil from Puglia made from hand picked
Coratina, Leccino and Olearola olives.
They blend green and black olives in equal
measure. It is a well made, tasty oil with a
clean astringent finish. Not assertive and
aggressive but unusually green for oil from
the south of Italy which may be due to the
proportion of green olives.
(LW)

AMABILE UMBRO

Appearance......Clear and bright

Colour......Yellow

Aroma......Ripely sweet and fruity

Taste......Light, soft, very delicate and sweet.
Quite a peppery finish

Uses......Pour over steamed vegetables, delicate fish,
salads and soups

Price range...... 🌱 🌱

COMMENTS
PDO oil from Umbria. This is a very attractive, delicate oil,
suiting fine flavoured dishes.
(A)

AZIENDA COLLI ALTI

Appearance......Clear and bright

Colour......Golden with a tinge of green

Aroma......Lovely enticing scent of green leaves and
cut grass with an astringent edge

Taste......Pleasant light style green bitter oil with a flavour
of apple and salad greens, just a hint of pepper

Uses......Raw on salad leaves, vegetables, fish,
pasta and for cooking

Price range...... 🌱

COMMENTS
Unusually light for a Tuscan oil. Well made, well balanced with just
the right amount of pepper on the finish. Almost Umbrian in style.
Very versatile and thoroughly recommended. Unbelievably cheap.
(FOC)

BADIA A COLTIBUONO

Appearance
Clear and bright

Colour
Golden

Aroma
Fresh tingle of crushed green grass
and bitter greens

Taste
Light but flavourful with a rasp of
pepper. Leaves a lovely silky taste of
roasted almonds

Uses
Salads, salad leaves, vegetables,
soups, stews, fish and grilled meat

Price range

COMMENTS
Estate oil of the highest quality from Tuscany,
made from hand picked Frantoio, Leccino
and Pendolino olives. Traditionally made, it is
an oil you can use anytime, anywhere,
versatile and delicious. Organic.
(OM)

GREECE

BAGGIOLINO LAUDEMIO

Appearance
Clear and bright

Colour
Dark golden

Aroma
Fresh, intense aroma of
green apple

Taste
Creamy in the mouth, rich and
nutty to start with but a fair
amount of pepper on the finish,
leaves a glorious green bitter taste

Uses
Raw on salad leaves, beans,
soups, bruschetta, pasta, dipping
oil, vegetables

Price range
Ψ Ψ Ψ

COMMENTS
Utterly delicious oil, bags of character
and a wonderfully structured, complex
array of flavours. I used to be a bit
sceptical about the Laudemio oils, and
in fact the ones I had previously tasted
did not impress and I felt they were just
fancy packaging; well this Tuscan
delight has converted me. Made from a
blend of Frantoio, Moraiolo, Leccino
and Pendolino, a classic Tuscan
cocktail.
(OM)

BADIA ALBERETO

Appearance......Clear and bright

Colour......Golden

Aroma......Bitter salad leaves and rocket

Taste......Round and creamy, quite a kick of pepper.
Very mild grassy flavour

Uses......Salad leaves, vegetables

Price range...... 🌱 🌱 🌱

COMMENTS
From the same producers as Badia a Coltibuono. Made from
Frantoio, Leccino, Moraiolo and Pendolino. Well made oil but a
bit too lacking in flavour and character for my taste. Badia a
Coltibuono is cheaper too.
(OM)

BARBERA FRANTOIA

Appearance......Clear and bright

Colour......Golden yellow

Aroma......Lovely fresh bitter green leaves

Taste......Light in style, delicate almond flavour with
a tickle of pepper

Uses......Salad leaves, vegetables, soup, grilled fish

Price range...... 🌱

COMMENTS
Lovely elegant, very delicate oil. No bitterness just smooth and
silky. Another fine example of Sicilian oil made from Biancolilla,
Nocellara and Cerasuola olive varieties.
(CAM) (OM)

ITALY

BAGGIOLIO

Appearance
Clear and bright

Colour
Golden

Aroma
Crushed green leaves

Taste
Very attractive astringent green bitter
flavour of salad leaves but a soft finish.
Almost no pepper in the throat but
pepper tingles on the tongue

Uses
Salad leaves, vegetables, pasta

Price range...... 🌱 🌱

COMMENTS
Lovely creamy, beautifully balanced well
structured oil, a delight to taste. From the
same estate as the Baggiolino Laudemio,
this is their second oil.
(OM)

BASILICA

Appearance
Hazy – unfiltered

Colour
Green gold

Aroma
Artichoke

Taste
Very fine elegant, delicate flavour of asparagus and artichoke

Uses
Dipping oil, bruschetta, vegetables, pasta, salads, fish and meat

Price range

COMMENTS
A truly aristocratic oil, so lithe and elegant. Very tantalising as the flavour lingers on the palate. Light and gentle style made from a mixture of Frantoio, Leccino and San Felicano olives in equal measure grown on the Tega Estate on the hills around the Basilica of St Francis of Assisi, hence the name of the oil. Certainly one to remember. I've got an open bottle in my kitchen now and I'm happily pouring it over everything. Amazing value.
(GFF)

BLUEBIRD

Appearance......Hazy – unfiltered

Colour......Golden

Aroma......Gorgeous fresh aroma of green leaves, rocket and artichoke

Taste......Creamy and smooth in the mouth with a sizeable kick of pepper. Light, delicate and bitter taste.

Uses......Let your imagination be your guide

Price range......

COMMENTS
Really superb oil. Very well balanced (there's quite a lot of pepper at the moment but that will mellow), elegant and characterful. Thoroughly recommended.
(P)

BOTTARELLI

Appearance......Clear and bright

Colour......Yellow

Aroma......Fresh green apple, green leaf and grass

Taste......Light, delicate but characterful. Light green tasting with a reasonable tickle of pepper on the finish. Leaves a nice astringent green leaf flavour lingering in the mouth

Uses......Raw on vegetables, fish and salads

Price range......

COMMENTS
This is a really excellent example of how a light delicate oil does not mean no taste or character. It is lithely elegant. Estate oil made from hand picked Leccino and Casaliva olives, grown on the hills near the western shore of Lake Garda.
(GFF)

CARAPELLI

Appearance......Clear and bright

Colour......Golden

Aroma......Sweet and ripely fruity

Taste......Mild olivey flavour with a little tickle of pepper

Uses......Cooking and salad dressings

Price range......

Not a bad blended Italian oil. You have to chew it a bit to get much flavour but it does leave a lingering taste of olive fruit.

CARLUCCIO'S LIGURIAN

Appearance......Clear and bright

Colour......Golden

Aroma......Quite intoxicating delicate green leaves

Taste......Light, delicate silky oil with a delicious little tingle of pepper on the tongue, leaving a wonderful lingering taste of green fruit and leaves

Uses......Dipping oil, salads, pasta

Price range......

COMMENTS
What a lovely oil. Quite seductive in the way it slowly reveals its flavour. I actually drank the entire tasting glass of this. Made from Taggiasca olives which is fast proving to be my favourite olive variety.

CARLUCCIO'S MARCHE

Appearance......Clear and bright

Colour......Yellow

Aroma......Crushed green leaves, bitter salad leaves and rocket

Taste......Fruity and green, finishes with an astringent edge of artichoke and sorrel

Uses......Vegetables, salads, salad leaves, grilled fish and meat

Price range...... ⍦ ⍦ ⍦

COMMENTS

Organic oil from near Ancona, made from hand picked Leccino, Frantoio, Maurino, Pendolino and Carboncella olives. This oil has grown on me. When I first tasted it I wasn't keen but it is such a complex oil, it takes time to reveal all its flavours. It is a wonderfully structured, green tasting oil.

CARLUCCIO'S PUGLIA

Appearance......Hazy-unfiltered

Colour......Green gold

Aroma......Delicious scent of banana skin and rocket leaf

Taste......Smooth, soft and silky, with a kick of pepper on the finish. Lightly fruity with an edge of mild artichoke.

Uses ......Raw on vegetables, fish, pasta

Price range...... ⍦ ⍦

COMMENTS

A lovely, soft fruity gentle oil, perfect where you have delicately flavoured food. Ideal condiment oil.

CARLUCCIO'S SARDINIAN

Appearance
Clear and bright

Colour
Yellow

Aroma
Truly gorgeous blast of freshly mown green grass

Taste
Delicately sharp flavour of green grass

Uses
Raw over salads, vegetables, dipping oil, grilled meat, vegetable soups

Price range
Ɏ Ɏ

COMMENTS
Made from Tonda di Cagliari, Bosana and Nera di Gosson olives, native varieties to Sardinia. This is a wonderful green oil which slides easily across the palate leaving a tingle of green leaf and a hint of asparagus.

CARLUCCIO'S SICILIAN

Appearance......Hazy – unfiltered

Colour......Golden

Aroma......Gorgeous intense aroma of fresh mown green grass

Taste......Smooth and silky in the mouth, delicious bitter green leaf taste

Uses......Pour it over everything

Price range......

COMMENTS
This is a lovely oil with bags of character. You really know you've got an olive oil when you use this. Buy a bottle and see.

CARLUCCIO'S UMBRIAN

Appearance......Clear and bright

Colour......Yellow gold

Taste......Packs a punch on the finish with a dose of pepper, smooth and silky on the tongue leaving a delicate flavour of almonds and a spicy prickle of pepper.

Uses......Pour over grilled meat, fish, pasta and robust soups and stews

Price range......

COMMENTS
I do like Umbrian oils, they are warm and spicy and this one is no exception. Made from hand picked Moraiolo, Leccino and Frantoio olives, this is a lovely example of oil with character.

CASSINI VITTORIO

Appearance......Hazy – unfiltered

Colour......Yellow

Aroma......Delicately spicy, green and bitter

Taste......Delicate, nutty taste, not at all bitter, lingering flavour of almonds

Uses......Vegetables, fish, soups

Price range...... ⚘

COMMENTS
A well made Ligurian oil but not really very exciting or characterful. There are better Ligurian oils around.

COL D'ORCIA

Appearance......Clear and bright

Colour......Yellow

Aroma......Fresh zingy aroma of salad greens and rocket

Taste......Green leaves and rocket

Uses......Drizzled over vegetables, meat, fish, soups, pizza, bruschetta and a good dipping oil

Price range...... ⚘ ⚘

COMMENTS
Lovely green oil which leaves a lingering astringent taste in the mouth. Just the right amount of pepper. Very versatile, can be used for many dishes. A classic Tuscan style oil.
(A)

COLONNA

Appearance
Clear and bright

Colour
Golden

Aroma
Ripe olives

Taste
Creamy and silky, with the mildest of peppery tickles. Very delicate taste of rocket leaves and finishes with a hint of almonds

Price range
Ψ Ψ

COMMENTS

I used to be very fond of Colonna but I have found myself less impressed in recent years, not because it's bad oil, far from it, but it seems to lack the character it had in the past. I think perhaps they have changed the blend or proportions of the olives used. They grow a few different varieties predominantly Coratina, mill them separately and then blend the resulting oils. What is however still excellent is their Granverde, an oil which has lemons pressed with the olives. I don't normally go for flavoured oils but this isn't flavoured it's a natural amalgamation. Utterly delicious over grilled fish.

(OM)

DEL SERO

Appearance......Clear and bright

Colour ......Golden

Aroma......Deliciously fresh and green

Taste......Light, soft oil, very delicate
flavour of artichoke with a
perfect level of pepper finish

Uses......Raw over salad leaves,
tomatoes, vegetables, fish
and soups

Price range......

COMMENTS
Very delicate, elegant oil from Umbria with a
whisper of flavour of cooked artichoke.
Excellent if you like finely tuned oils.
(OG)

FILIPPO BERIO ORGANIC

Appearance
Clear and bright

Colour
Golden

Aroma
Bright aroma of crushed bitter leaves with a tickle of pepper

Taste
Light flavour of olive fruit, leaves and a finish of almonds

Uses
Salad leaves, salad dressings, vegetables

Price range

COMMENTS
Good balanced attractive style. If you want an everyday affordable organic oil, this is probably it. This is the best of the Berio range. The ordinary EV is good value, has flavour and a nutty finish but the Special Selection is tasteless and characterless and definitely not worth buying.

COLLINE DI MASSA MARITTIMA

Appearance......Clear and bright

Colour......Yellow

Aroma......Ripe crushed olives with a tinge of pepper

Taste......Bitter green leaves

Uses......Raw on bruschetta, vegetables and soup

Price range......

COMMENTS
A very well balanced harmonious, medium weight Tuscan oil.
Tasty and green but not too bitter and very little pepper
on the finish.
(A)

DONNAVASCIA

Appearance......Clear and bright

Colour......Yellow

Aroma......Lovely immediate noseful of sliced apple

Taste......Light, delicate but no flavour

Uses......Salad dressings

Price range......

COMMENTS
After the delicious aroma I was expecting a mouthful of green
apple but the oil has really no flavour or character.Disappointing
estate oil from Calabria made from
Carolea olives.
(GFF)

FIOR FIORE

Appearance......Hazy – unfiltered

Colour......Yellow gold

Aroma......Sweet, almost honeyed

Taste......Soft, gentle and nutty, very tiny tickle of pepper

Uses......Vegetables, soups

Price range......

COMMENTS
Deliciously easy tasting oil from Puglia. Truly amazing value for money. How do they do such good oil at this price?
(V&C)

FIORE DI FRASCHIERA

Appearance......Clear and bright

Colour......Yellow

Aroma......Light bitter green aroma with an appley edge

Taste......Soft gentle mouthful, developing quite a peppery finish

Uses......Salad dressings and cooking

Price range......

COMMENTS
The label says this is a full-bodied oil but it doesn't have much flavour at all and leaves nothing behind on the palate. Disappointing.
(A)

FRANTOIO SANTA TEA INTENSO

Appearance
Clear and bright

Colour
Green gold

Aroma
Intense, fresh bitter green leaves
and green apple

Taste
Wonderful, smooth green bitter oil but light in
style leaving just a tantalizing, lingering flavour
of green unripe banana and a tingle of
pepper on the tongue but not on the finish

Uses
Pour raw over vegetables, salads, grilled
meats, soups and casseroles, as a dipping oil
and on bruschetta

Price range

COMMENTS
Such lovely oil, always consistently good. Not at
all peppery on the finish, harmonious,
well balanced, in fact everything top class oil
should be. From an estate not far from Florence,
made from Carolea olives. The Intenso is made
from olives harvested early, while still green and
the label shows green olives, while the Delicato,
which is sweeter and softer, is made from olives
harvested when fully ripe. The label for Delicato
shows a red olive. Both are excellent and rate
among my favourite Italian oils.
(GFF)

ITALY

FRESCOBALDI LAUDEMIO

Appearance......Clear and bright

Colour......Golden

Aroma......Amazing scent of green herbaceous and mustard greens

Taste......Smooth, round rich but soft, flavour of green leaves and a perfect level of pepper on the finish

Uses......Pasta, beans, vegetables, soups, casseroles, salads, dipping oil, bruschetta

Price range...... ⅄ ⅄ ⅄

COMMENTS
Really very attractive immediate intense oil made from Frantoio, Moraiolo and Leccino. The tongue tingles with flavour, pepper and greenness.
(OM)

GAZIELLO VENTIMIGLIA

Appearance......Clear and bright

Colour......Yellow

Aroma......Lovely aroma of grated green apple

Taste......Thick, smooth and silky in the mouth but the flavour is light and stylish. It is nutty, reminiscent of almonds and vanilla

Uses......Raw on vegetables, salad leaves, soups

Price range...... ⅄ ⅄

COMMENTS
This oil is made from Taggiasca olives, the most common variety in Liguria. It is utterly delicious when the oil is first made so best to buy when it's very new. Very nice indeed, worth trying.
(FOC)

IL CARDINALE

Appearance......Hazy – unfiltered

Colour......Green gold

Aroma......Full, pungent aroma of ripe crushed olives, with a bitter green edge

Taste......Very peppery, very bitter green finish

Uses......Soups, casseroles, grilled meats

Price range......

COMMENTS
Oil from Puglia. Looking at the glorious colour and inhaling the wonderful aroma I thought I was in for a treat but the oil never really gets going and doesn't have any distinctive flavour. Disappointing.
(FOC)

IL FATTORE

Appearance......Clear and bright

Colour......Golden yellow

Aroma......Delightfully perfumed warm almonds

Taste......Light in style with a delicate caramalized nutty, fruity finish. No pepper at all

Uses......Raw on vegetables, salads and fish

Price range......

COMMENTS
A lovely light delicate oil from Perugia but with taste and character.

IL NUMERATO RIVIERA LIGURE

Appearance......Clear and bright

Colour......Golden

Aroma......Sweet aroma of sun warmed fruit

Taste......Light, bright bouncy oil, delicate sweet flavour of nuts and a touch of green leaf

Uses......Cooking and salad dressings

Price range......

COMMENTS
PDO oil bottled by Carapelli. A very good example of typical Ligurian style oil.

IL NUMERATO TOSCANO

Appearance......Clear and bright

Colour......Greenish gold

Aroma......Green leaves with a very pleasant bitter edge

Taste......Soft, smooth flavour of green leaves and artichoke with a good tickle of pepper

Uses......Cooking, salad dressings, grilled meat

Price range......

COMMENTS
This is a PGI oil bottled by Carapelli. An excellent example of a commercially blended EV made from a blend of Frantoio, Leccino and Moraiolo. It's got a good flavour and some character, it doesn't leave much behind on the palate but thank goodness it is possible to determine typical Tuscan style in this (shows there is some purpose to designations such as PGI). I also commend them for putting harvest year on the neck band.

IL NUMERATO UMBRIA

Appearance......Clear and bright

Colour......Greenish gold

Aroma......Bitter greens and pepper

Taste......Round and smooth with a hefty kick of pepper on the finish. Taste of caramalized almonds

Uses......Cooking and salad dressings

Price range......

COMMENTS
PDO oil bottled by Carapelli, says on the label it's from Colli Assisi-Spoleto. A very good classic example of an Umbrian style oil made from Moraiolo olives. Medium weight, with deep chocolatey notes, flavoursome and characterful. I like the fact they put the harvest date on the bottle.

IL PADRONE

Appearance......Clear and bright

Colour......Greenish gold

Aroma......Heady perfume of green leaves, crushed bitter greens and rocket

Taste......Medium weight, round and ultra silky, very mild flavour of sorrel with a tickle of pepper on the finish. Sweeter rather than bitter

Uses......Raw on fish, and vegetables and as a dipping oil

Price range......

COMMENTS
This is a PGI oil from Tuscany. Much more Umbrian in style though, it has a deep fruity flavour but is not at all heavy. Made from early harvested Leccino and Moraiolo olives but interestingly not green to the taste as you might expect from early harvest oil. Ideal for those who like very mild flavoured oils.
(CAR)

LA COLLINA

Appearance......Clear and bright

Colour......Dark greenish gold

Aroma......Faint tang of artichoke

Taste......Light in style, quite a peppery finish, very subtle taste of cut artichoke

Uses......Raw on salad leaves and vegetables

Price range......

COMMENTS
Estate oil from the Tega family mill made exclusively from Moraiolo olives grown in the Assisi region of Umbria. Very subtle, quite elegant oil but finishes just a bit too quickly for my taste, could do with a little more character.
(GFF)

LA MOLA

Appearance......Clear and bright

Colour......Golden

Aroma......Grassy

Taste......Delicate green sorrel leaves

Uses......Vegetables, soups, grilled meat

Price range......

COMMENTS
Oil from Latium made from Frantoio and Leccino olives. Lovely and round and smooth in the mouth. Quality oil from www.esperya.com.

LA RENA

Appearance......Clear and bright

Colour......Golden

Aroma......Ripe, crushed olives

Taste......Quite a peppery kick but rather flat,
dull fruity flavour

Uses......Cooking

Price range......

COMMENTS
Made from hand picked Ogliarola, Cellina and Coratina olives.
Sadly though not a very exciting oil at all. Lacks any character
and leaves no taste.
(A)

LE FASCE

Appearance......Clear and bright

Colour......Yellow gold

Aroma......Peppery tickle on the nose and an
aroma of warm hay

Taste......Fades very fast, not much flavour

Uses......Salad dressings

Price range......

COMMENTS
It's well made but lacks any flavour or character.
(GFF)

LE TREBBIANE

Appearance......Clear and bright

Colour......Golden

Aroma......Delicate green leaves

Taste......Round, creamy, pepper finish.
Delicate green bitter taste

Uses......Salad leaves, vegetables, soups,
dipping oil, bruschetta

Price range......

COMMENTS
Very nice oil indeed from Tuscany. Elegantly bitter and delicious.
Kind of oil you can use and use because of its easy style.
(EN)

LILLIANO

Appearance......Clear and bright

Colour......Dark green gold

Aroma......Crushed green leaves and apples

Taste......Light and silky smooth in the mouth. Lemony to
start, finishes with the right shot of pepper.
Leaves a lovely subtle green leaf taste in
the mouth.

Uses......Salad leaves, grilled fish, soups and salads

Price range......

COMMENTS
Very harmonious and balanced oil. Classic example
of a Tuscan oil.
(OM)

LUNGAROTTI CANTICO

Appearance
Clear and bright

Colour
Yellowish green

Aroma
Very perfumed and aromatic,
reminiscent of warm hay

Taste
Bitter green leaves and artichoke.
Amazing flavour, it bursts on the palate
almost with a fizz

Uses
Pour over vegetables, meat,
casseroles, beans, pasta, salads

Price range
Ψ Ψ

COMMENTS
What a sensational and unique Umbrian oil.
It's harmonious, well balanced, delicious.
You could almost sit and sip it from a glass.
I have enjoyed every Lungarotti oil I have
tasted. They are first class, prize winning
producers who really know how to make oil
of the highest quality.
(A)

MANDRANOVA

Appearance......Hazy – unfiltered

Colour......Green gold

Aroma......Cut green leaves and green banana

Taste......Very delicate bitter green

Uses......Salads, salad leaves, vegetables, grilled meat, fish, beans, bruschetta

Price range...... ❦ ❦

COMMENTS
Very well balanced, harmonious deliciously delicate bitter green oil. Another fine example of the class of Sicilian oils.
(EN)

MARCHESI DI PIETRAFORTE

Appearance......Hazy – unfiltered

Colour......Green gold

Aroma......Fresh cut green grass

Taste......Lightly nutty, leaves a tickle of pepper on the tongue

Uses......Salad leaves, soups, grilled fish

Price range...... ❦

COMMENTS
Delicate Umbrian oil which leaves a pleasant hint of almonds. What makes this oil doubly attractive is the ingenious packaging allowing you to buy it loose. The places where it is sold keep these stylish pouches and fill them according to need when you buy. No more having to remember to take a bottle when you want to buy loose oil. Such a simple solution and one I'm sure many other people will follow in due course.
(OG)

MONIGA DEL GARDA

Appearance......Clear and bright

Colour......Yellow

Aroma......Hint of sweetness and leaves

Taste......Round and smooth, creamy and silky, lovely flavour of almonds mixed with olive fruit

Uses......Vegetables, salads and salad leaves, fish, soups

Price range......

COMMENTS
Sweet style oil, deliciously smooth with medium weight.
(EW)

OILBIOS

Appearance......Clear and bright

Colour......Yellow

Aroma......Mild aroma of green leaves and a hint of artichoke

Taste......Lovely delicate bitter flavour of salad leaves and rocket, with just a modest tingle of pepper on the finish

Uses......Pour over salad leaves, vegetables, soups

Price range......

COMMENTS
Great organic oil from hand picked olives. This really is subtle, delicate, well structured and harmonious. Use it where you want to enhance the flavour of food, it's a perfect foil. Very good value. This is just one of the Umbrian oils from Monini. All the oils are soundly good and they deserve to be much more widely stocked.
(A)

MONTE VERTINE

Appearance
Clear and bright

Colour
Dark gold

Aroma
Faint green leaf

Taste
Creamy, lightly fruity with a little
edge of green

Uses
Raw over vegetables, salads, grilled
meat, fish and as a dipping oil

Price range

COMMENTS
Tuscan estate oil from Chianti made from
Correggiolo, Moraiolo and Leccino olives.
The bottle I tasted was from the 1998 harvest.
This oil when it's newly pressed is vivid green in
colour and taste. This sample being over 18
months old has mellowed and softened but it
is still showing without difficulty that it is a true
class oil.
(EW)

MONTIFERRU

Appearance
Slightly hazy

Colour
Yellow gold

Aroma
Ripe fruity olives

Taste
Sweet and nutty, flavour of almonds,
tickle of pepper

Uses
Salads, fish

Price range

COMMENTS
I can't believe the price of this oil. It is the
same price as the average supermarket oil.
Taste this and then understand why I
thunder against the appalling quality of
supermarket oils. If a company like
www.esperya.com can provide such a
delightful, sweet delicate Sardinian oil at this
price, then why can't everyone?

NATIVO ORGANICO

Appearance......Clear and bright

Colour......Yellow

Aroma......Rich fruity nose with hints of crushed
green leaves

Taste......Fills the mouth with very delicate fruit but
then is gone leaving nothing behind

Uses......Steamed vegetables and fish

Price range......ᵞ ᵞ

COMMENTS
Organic oil, estate grown and pressed, from Campania. Too
fleeting a taste to be a truly satisfying oil. It is obviously well made
but it just doesn't have enough taste or character.
(GFF)

OLIVES ET AL CORATINA

Appearance......Clear and bright

Colour......Golden yellow

Aroma......Overripe crushed olives

Taste......Round and rich, with a mild peppery tickle on
the finish. A bit one dimensional

Uses......Salad dressings, cooking

Price range......ᵞ ᵞ

COMMENTS
An oil from Apulia in the south of Italy, from a single variety olive,
Coratina. This is not the best example of Coratina I have tasted. It's
just not very exciting.
(OEA)

OLIVES ET AL OGLIOROLA DI MORA

Appearance......Clear and bright

Colour......Golden yellow

Aroma......Sweet and fruity

Taste......Light style. Sweetly fruity. Quite a powerful burst of pepper on the finish. Deep chocolatey flavour but delicately so

Uses......Raw on salads, vegetables, soups, beans, pasta

Price range......

COMMENTS
Oil from a single olive variety probably little known over here, common to Apulia. Very low acidity, 0.3%. This is a very attractive oil indeed, good balance, pleasantly fruity, versatile.
Well worth a try.
(OEA)

PETRAIA

Appearance......Clear and bright

Colour......Golden

Aroma......Floral and grassy

Taste......Light in style, nutty with a lingering flavour of bitter leaves, quite a kick of pepper

Uses......Salads, salad leaves, soups, vegetables

Price range......

COMMENTS
A well made delicate organic oil mainly from Coratina olives but not very distinctive or different.
(OM)

PIETRANTICA

Appearance......Clear and bright

Colour......Golden

Aroma......Lovely lush grassy aroma with grated apple

Taste......Light style and very delicate nutty flavour

Uses......Raw on fish and vegetables

Price range......

COMMENTS
Well made, harmonious oil from Liguria, for those who like
very delicate oils. It is exceptionally good value for an oil
of this quality.
(GFF)

PODERE COGNO

Appearance......Clear and bright

Colour......Gold

Aroma......Fresh grassy aroma

Taste......Lovely, silky, smooth oil with a tickle of pepper.
Nutty, deep chocolatey notes with touches
of dried fruit

Uses......Salads, dipping oil, bruschetta, vegetables

Price range......

COMMENTS
This is an excellent estate oil from Tuscany made from Frantoio,
Leccino and Moraiolo olives. Elegant. Recommended.
Exclusive to Harvey Nichols.

POGGIO LAMENTANO

Appearance
Hazy – unfiltered

Colour
Gold

Aroma
Gorgeous rush of bitter green leaves,
almost perfumed

Taste
Medium weight, good tickle of pepper,
softly green with a bitter finish

Uses
Salads, salad leaves, dipping oil,
vegetables, soups, stews, grilled
meat or fish

Price range
Ψ Ψ Ψ

COMMENTS
This oil comes from an estate owned by the
Zyw family who moved to Tuscany in the
1960s. It was a favourite of Elizabeth
David's; she sold it and championed it when
she had her shop in London in the 70s.
A truly exceptional quality Tuscan oil made
from predominantly Moraiolo, with some
Frantoio, Leccino and Pendolino. Classically
green in aroma and taste. They always put
the date of harvest on the label.
(V&C)

ITALY

RAVIDA

Appearance
Slightly hazy – unfiltered

Colour
Green gold

Aroma
Big on the nose with crushed green leaves and green banana and end note of tomato

Taste
Tomato, green leaves and apple

Uses
On everything you eat

Price range
Ψ Ψ

COMMENTS
What can I say that hasn't already been said about this true aristocrat among oils? It is one of the most complex oils both in aroma and flavour, revealing itself little by little. Made from olives unique to the area of Sicily: Nocellara del Belice, Biancolilla and Cerasuola. If you want to understand what great olive oil is about then buy this and you have an instant lesson in one bottle. It's cheap for such excellence.
(OM)

RAGGIO DI SAN VITO

Appearance......Clear and bright

Colour......Golden yellow

Aroma......Perfumed, an amazing aroma of almonds

Taste......Light and nutty, flavour of sweet almonds

Uses......Salads, soups, grilled fish, vegetables
and over sliced oranges

Price range...... ⚊ ⚊

COMMENTS
Quite an extraordinary flavour, unlike any other oil I have
tasted. It's light and fruity but finishes with a distinct taste of
honey and almonds. It's elegant and well worth a try for the
novelty of the taste.
(CAR)

RIVER CAFE CAPEZZANA

Appearance......Clear and bright

Colour......Deep gold

Aroma......Delicate green bitter leaves

Taste......Very light and soft in style, spicy flavour
with a lingering green bitter finish

Uses......Vegetables, grilled fish, salad leaves

Price range...... ⚊ ⚊ ⚊

COMMENTS
Ancient estate of Tenuta Capezzana north of Florence
produces this blend of predominantly Moraiolo with some
Frantoio olives. It is subtle, delicate, elegant and nutty,
almost a taste of sweet almonds but leaves a gentle peppery
coating over the palate.
(OM) (LW)

RIVER CAFE SELVAPIANA

Appearance
Clear and bright

Colour
Golden

Aroma
Rich and green and fruity

Taste
Bitter green leaves and rocket, spicy,
peppery tickle on the finish

Uses
Pasta, salad leaves, grilled meat and fish,
dipping oil, bruschetta, vegetables, soups,
casseroles

Price range

COMMENTS
Single varietal oil made from Frantoio olives grown
on the Selvapiana estate, north of Florence. It's
lovely and bitter almost mouth-puckeringly so, but
medium weight. Leaves a real tingle of chilli
pepper on the palate. Excellent stylish Tuscan.
There is another River Cafe oil in the range, Fatoria
di Morello, also from near Florence but made from
Moraiolo olives. It is a very characterful classic
Tuscan, bitter green oil. Lovely.
(LW)

RIVER CAFE I CANONICI

Appearance......Hazy

Colour......Gold

Aroma......Intense fresh rocket leaves and pepper

Taste......Round and creamy in the mouth, quite
strong pepper, flavour of green and bitter
leaves, slightly spicy

Uses......Pasta, vegetables, salad leaves, casseroles,
grilled fish and meat, robust dishes,
dipping oil

Price range......

COMMENTS
Very green bitter oil, intense, full flavoured and characterful,
made from Moraiolo olives grown south of Florence. Excellent.

ROI

Appearance......Hazy – unfiltered

Colour ......Golden

Aroma......Ripe and fruity, vanilla and honeyed nuts

Taste......Soft, elegant and light on the tongue.
Deep nutty chocolatey notes

Uses......Salad leaves, salads, soups, grilled fish,
steamed vegetables

Price range......

COMMENTS
A truly delightful oil produced by the Du Roi family from
Taggiasca olives grown near Badalucco in the Argentina Valley,
Liguria. You do need to look for a long best before date on
these oils though because Taggiasca is, in my opinion, past its
best after six months from pressing. Thoroughly recommend
this and their free run Carte Noir oil.
(D)

SEGGIANO

Appearance
Hazy – unfiltered

Colour
Golden

Aroma
Warm hay

Taste
Silky and creamy in the mouth, delicate flavour of nuts and almonds

Uses
Salad leaves, soups, vegetables, dipping oil, grilled fish, pasta

Price range

COMMENTS
It's very clear why this oil is such a hit with so many people and is stocked so widely; it's easy, light and delicately flavoured. A most unique Tuscan oil, not at all green, assertive or bitter. The original flavour comes from an olive variety native only to this village, Olivastra Seggianese. It is a single estate, organic single varietal oil made from hand picked olives traditionally pressed. I find it heartening that oil made by one couple, David Harrison and Peri Eagleton, which started as a hobby but then out of necessity became a business, has been so successful. (P)

SERÉGO ALIGHIERI

Appearance......Clear and bright

Colour......Golden yellow

Aroma......Upfront and very fruity, ripe olives and hay

Taste......Rather flat, tired, not much flavour

Uses......Salad dressings

Price range......

COMMENTS

The Alighieri family are descendants of the poet Dante. This oil made from hand picked, stone crushed olives, comes from their estate in Valpolicella. Rather disappointing. There was no 'best before' date on the bottle I was sent from the distributor, so it's difficult to judge if it's old stock. It certainly tasted like oil past its best. However my request for another bottle was met with silence so I can't give a real opinion. Stocked as part of Sainsbury's Special Selection.
(E)

RUSTICO DI SICILIA

Appearance......Clear and bright

Colour......Golden yellow

Aroma......Zingy tang of chopped green leaves

Taste......Soft and silky in the mouth which then gives the merest tickle of pepper, leaving a delicate flavour of green fruit

Uses......Raw on vegetables, salad leaves, fish, soups

Price range......

COMMENTS

Sicily certainly does manage to provide consistently lovely, stylish, characterful oils and this is no exception. Amazing value, almost the price of supermarket oil with a thousand times more class.
(GFF)

TENUTA DI SARAGANO

Appearance
Clear

Colour
Golden

Aroma
Fruity olive

Taste
Lovely medium weight oil, round and soft in the mouth, just a tingle of pepper on the finish. Flavour of delicate toasted nuts

Uses
Raw on salad leaves, vegetables, fish and meat

Price range

COMMENTS
This is a delicate fine structured estate oil from Umbria. A benchmark for quality, showing how you can have character, flavour and elegance but with a light style. A truly delicious oil made from a combination of Frantoio, Leccino and Moraiolo olives. This has long been a real favourite oil of mine and I think it's excellent value.
(OM)

TERRICCI

Appearance......Clear and bright

Colour......Golden

Aroma......Delicious, immediate, intense green apple

Taste......Soft and sweet start which develops into a lovely bitter green leaf flavour finishing with a tickle of pepper

Uses......Raw on pasta, vegetables, meat, fish, soups and casseroles

Price range......

COMMENTS
Tuscan, unfiltered oil. Light in style but not lacking in flavour. (WW)

VANINI

Appearance......Clear and bright

Colour......Golden yellow

Aroma......Hint of almond but a bit flat

Taste......Light flavour of almonds, with a pleasant level of pepper on the finish

Uses......Salads and vegetables

Price range......

COMMENTS
Hand picked, stone crushed oil from Lake Como. Very low acidity, 0.2%, well made and well balanced but lacking any real character. (BDV)

TIBERIO ABRUZZO

Appearance
Hazy – unfiltered

Colour
Golden yellow

Aroma
Sweet and ripely fruity, sun-warmed olives

Taste
Light style, sweet and fruity with a finish
of almonds

Uses
Salad leaves, vegetables, pasta

Price range

COMMENTS
The more south you go in Italy the sweeter the
oils. This one from the Abruzzo is pleasantly
delicate in taste and light in style. It is made from
an olive variety called Gentile di Chieti, which is
derived from the Frantoio. Delightful oil.
(V&C)

U TRAPPITU INTENSO

Appearance
Slighly hazy

Colour
Green gold

Aroma
Absolutely delicious aroma of cut grass

Taste
Light on the tongue, tickle of pepper.
A mouthful of the most wonderful
tastiest bitter green leaves and green
apple. But in truth any words of
description fail to conjure up the sheer
beauty of the flavour of this oil

Uses
On everything

Price range
Ψ Ψ

Comments
This is so utterly gorgeous. It's elegant, lithe,
green without any hint of bitterness. Sicily
certainly seems to be blessed with the finest
oils it has been my pleasure to taste in the
last two years. This is made from that divine
triumvirate of Biancolilla, Cerasuola and
Nocellara del Belice olives. Perfection in a
bottle. There is also a Delicato, which is
equally delicious. Both from
www.esperya.com

ITALY

VALGIANO

Appearance
Clear and bright

Colour
Golden yellow

Aroma
Amazing scent of zested citrus

Taste
Nutty and fruity, round in the mouth with
a peppery finish

Uses
Salads, grilled meat, dipping oil, bruschetta

Price range
Ψ Ψ Ψ

COMMENTS
An attractive Tuscan oil from Lucca with
character and flavour, it prickles the tongue with
peppery notes but has a background flavour that
is fruity but light.
(OM)

TENUTA DI VALGIANO
Olio
Extra Vergine
di Oliva

Messo in bottiglia dalla
TENUTA DI VALGIANO

Valgiano Lucca Italia

500 ml e 17 fl. oz.

Lotto N. 1299 Raccolto Novembre 1999

Da consumarsi preferibilmente entro MAGGIO 20
NON DISPERDERE IL VETRO NELL'AMBIENTE

CONSERVAS RAINHA SANTA ORGANIC

Appearance
Hazy – unfiltered

Colour
Golden

Aroma
Ripely fruity, rustic and vegetal

Taste
Very ripely fruity, taste of ripe olives

Uses
Soups, casseroles, dipping oil, robust dishes and ones with lots of garlic, grilled meat

Price range

COMMENTS
Surely a benchmark oil for the new era of Portuguese oils. This is in the classic style but much lighter and more delicate. Whereas in the past I haven't liked the earthiness of Portuguese oils (they were made of over ripe olives) this one is very attractive. What is exciting about this oil is that it has improved on the style without losing the unique character that says 'this is Portuguese oil'. An organic oil from Tras os Montes in the north of Portugal. Made from Galega, Alto Douro and Picoal olives. Delicious, do try it for a change. Conservas Rainha Santa have other oils in their range: an estate oil from Alentejo, which is sweet tasting with lots of character, and an elegant oil from Estremoz.
(CRS)

PORTUGAL

PORTUGAL

QUINTA DAS MARVALHAS

Appearance
Clear and bright

Colour
Dark gold

Aroma
Perfumed, very floral with green olive

Taste
Round, smooth and creamy, pepper on the finish. Elegant flavour of artichoke and bitter green leaves

Uses
Vegetables, salads, grilled fish, beans

Price range

COMMENTS
One of the first new-style Portuguese oils, produced by the Madeira family (see Portugal in country guide) from varieties native to Portugal. It is a PDO oil, delicate with quite a peppery finish. Well made and harmonious with enough flavour to be characterful. Nice all purpose oil and worth a try if you like to experiment with different tastes and styles.
(RR)

MORGENSTER

Appearance
Clear and bright

Colour
Yellow

Aroma
Bitter green leaf and pepper with a hint
of artichoke

Taste
Round and creamy, astringent warm
bitter green flavour

Uses
Grilled meat, fish, salad leaves

Price range
💰 💰

COMMENTS
An interesting and unusual oil, it's South
African. Not masses of character or flavour
but worth trying for a different style of oil.
(OM)

BLAZQUEZ

Appearance......Hazy – unfiltered

Colour......Dark gold

Aroma......Lovely intense aroma of grated apple

Taste......Round and smooth in the mouth, tingle of pepper on the back of the throat. Medium weight leaves flavour of green leaf and a hint of apple.

Uses......Vegetables, fish, bean dishes, salad dressings, cooking

Price range...... 🌱 🌱

COMMENTS
A well-made oil easy in style, made from Lechin olives grown south of Seville. Fine if you like very mild oil but I find the taste disappears a little too quickly.
(B) (OM)

BORGES

Appearance......Clear and bright

Colour......Golden

Aroma......Sweet, fruity and olivey with a hint of apple

Taste......Sweet, light and fruity, with a distinctive Spanish flavour, leaves a pleasant lingering aftertaste

Uses......Everyday cooking, salad dressings, soups and casserols

Price range...... 🌱

Comments
Blended oil from Catalonia in North Eastern Spain inland from Barcelona. This used to be made from Hojiblanca and Cornicabra, the present label suggests olives sourced from the North. This is a really well made oil. It has flavour and balance. Availability from supermarkets means it's widely accessible.
(AO)

BORGES ARBEQUINA

Appearance......Clear and bright

Colour......Golden yellow

Aroma......Green olive and green leaf with a hint of almond, fruity with a bitter edge

Taste......Smooth, lightly fruity and elegant

Uses......Drizzled over salad leaves, cooked vegetables and fish

Price range......

COMMENTS

This is a very good oil indeed. An excellent example of a mass-produced oil showing that production in quantity does not have to sacrifice quality. I like the fact that Borges print the bottling date on their labels as well as the best before date.
(AO)

BORGES HOJIBLANCA

Appearance......Clear and bright

Colour......Golden yellow

Aroma......Sweet and fruity, with just a bitter edge

Taste......Sweet and fruity but light in style

Uses......Salads, dressings, anywhere you want a distinctive flavour

Price range......

COMMENTS

A good quality oil for those who like this kind of flavour.
(AO)

SPAIN

BORGES PICUAL

Appearance......Clear and bright

Colour......Golden yellow

Aroma......Intense, sweet fruity aroma. Lovely tropical
fruit and banana skin

Taste......Fruity with a bitter edge

Uses......To give flavour to cooking, vegetable or
meat sauces, also good for drizzling over
toasted bread with garlic and tomato and
on potatoes

Price range...... ⅄

COMMENTS
A lovely distinctive oil with bags of flavour, useful for cooking
most Mediterranean style dishes.

BRINDISA ORGANIC

Appearance......Clear and bright

Colour......Golden yellow

Aroma......Ripe, warm olive fruit with a hint of bitter
green leaf

Taste......Round and full in the mouth, tingle of pepper
on the finish but very little flavour to catch
hold of

Uses......Salad dressings, grilled fish

Price range...... ⅄

COMMENTS
A bit too subtle in flavour.

CARBONELL ORGANIC 1866

Appearance
Clear and bright

Colour
Deep gold

Aroma
Lightly sweet and fruity, melon and banana

Taste
Light, fruity and sweet with just a tickle of pepper on the finish

Uses
Raw and for cooking

Price range ⑂

COMMENTS
Excellent example of well made, well balanced harmonious oil. Lighter and more elegant than the ordinary Carbonell EV. Fine organic oil, very adaptable and versatile. If you want to buy organic look no further. (C)

SPAIN

CARBONELL

Appearance
Clear and bright

Colour
Golden

Aroma
Sweet, of ripe melon and
banana skin

Taste
Sweet, fruity but not too heavy. Leaves
lingering taste of tropical fruit

Uses
Cooking for robust dishes or raw on
salads,vegetables and especially
baked potatoes

Price range

COMMENTS
Carbonell never fails to please. It is such a
good example of lush fruity Spanish oil
produced in bulk but never compromising
on quality and character. It is a benchmark
commercial blend southern Spanish oil.
They also produce a more expensive Special
Selection, which is very different in style. It
tastes of bitter greens and rocket, very
attractive astringent flavour.

FRESH OLIVE COMPANY – ARBEQUINA

Appearance......Hazy – unfiltered

Colour......Yellow

Aroma......Lovely immediate sweet, fruity aroma almost like a fruit jelly, with a tang of warm citrus fruit, perhaps grapefruit

Taste......Light in style, hint of pepper on the finish, lovely creamy sensation in the mouth. Delicate pleasant after taste, almost of sweet citrus

Uses......Cooking, raw on salad leaves, vegetables, grilled fish

Price range...... ⍙

COMMENTS
The Fresh Olive Company does have the knack of sourcing excellent oils which deserve to be as well known as their olives.

L'ESTORNELL ORGANIC

Appearance......Clear and bright

Colour......Golden yellow

Aroma......Ripe olives

Taste......Delicate almonds

Uses......Vegetables, soups and salads

Price range...... ⍙ ⍙

COMMENTS
Stylish, straightforward estate oil from Catalonia. Made from Arbequina olives. Always reliable. (OM)

COOPERATIVA CAMBRILS

Appearance
Clear and bright

Colour
Yellow

Aroma
Delicious immediate, sweet fruity aroma of olive and banana

Taste
Lovely smooth sweet, nutty oil. Leaves a very pleasant lingering taste with the tiniest tingle of pepper

Uses
Raw on salads, vegetables, fish, meat, soups and for cooking

Price range

COMMENTS
This is a delightful PDO oil from Siurana in North Eastern Spain, with an amazing acidity of 0.3%. It is well balanced, harmonious. Very easy and versatile. I would use it for everything, raw and cooking.
(B)

LERIDA

Appearance......Clear and bright

Colour......Yellow gold

Aroma......Perfumed with the warm aroma of almonds

Taste......Round and warm and smooth nutty flavour with just a tickle of pepper to finish

Uses......Salads, salad leaves, vegetables and fish

Price range......

COMMENTS
Estate oil from the north of Spain made from hand picked Arbequina olives. I visited the estate many years ago when I first started writing about olive oil and this oil from the Vea family is still delicious, well made and stylish. It is elegant and thoroughly recommended. Very reasonably priced too for an oil of such quality.
(OM)

MAS PORTELL ORGANIC

Appearance......Clear and bright

Colour......Golden

Aroma......Sweet, luscious and fruity, almost floral in its sweetness

Taste......Very light style, leaving just a residual sweetness lingering delightfully on the palate

Uses......Raw on salad leaves, vegetables, grilled fish, toasted bread, dipping oil

Price range......

COMMENTS
This is a most elegant, stylish Spanish oil from near Lleida, made from Arbequina olives and certified organic. Very light in style but doesn't lack flavour. Excellent value too for organic.
(FOC)

NUÑEZ DE PRADO

Appearance......Hazy – unfiltered

Colour......Yellow

Aroma......Intense, sweet, warm lemony, citrus with grated apple

Taste......Smooth and creamy, taste of bitter green leaf,very bitter finish with a touch of pepper

Uses......Vegetables, salads, salad leaves, grilled meat, grilled fish

Price range

Ψ Ψ

COMMENTS

Nuñez used to be my number one favourite Spanish oil. It packed a huge gob-smacking punch of tropical fruit on the palate and it was like inhaling an exotic fruit salad. It is still an extremely fine oil, one of the best but I have confess to not enjoying the style as much. I think they have changed the balance of the blend, which is made from Picual, Picudo and Hojiblanca olives. It is now a lighter oil which may appeal to a wider market but not so much to me sadly. (B) (OM)

MERIDIAN SPANISH ORGANIC

Appearance......Clear and bright

Colour......Yellow

Aroma......Very ripe olive fruit

Taste......Light style, but ripely sweetly fruity

Uses......Casseroles, soups and robust dishes

Price range......

COMMENTS
Quite a typical example of southern Spanish oil, this comes from near Cordoba and is made from Hojiblanca olives. It is Soil Association Certified and I can only assume this is why it has dominated the organic market for so long. It is for example the only oil stocked by Holland and Barrett. It is much improved from the samples tasted in the past, when frankly the oil was unspeakable but I do feel that there are very many better tasting organic oils now available.

SPAIN

OLIVES ET AL ARBEQUINA

Appearance......Clear and bright

Colour......Golden yellow

Aroma......Rather closed. Very faint sweetly fruity and a touch of cheese

Taste......Very light oil on the tongue. Hardly any lingering flavour, pepper on the finish

Uses......Drizzled over fish

Price range......

COMMENTS
This is a PDO oil from Les Garrigues in northern Spain, though sadly it has none of the distinctive characteristics I would expect from a PDO single varietal (Arbequina) oil. Rather disappointing. There are better Arbequina oils around.
(OEA)

OLIVES ET AL HOJIBLANCA

Appearance......Clear and bright

Colour......Golden yellow

Aroma......Sweet and fruity

Taste......Sweet and light in style

Uses......Pour over cooked vegetables and fish

Price range......

COMMENTS
The Hojiblanca olive variety is found growing in Andalucia in southern Spain, where it usually produces fruity oil but this is very light indeed, probably too light for my taste.
(OEA)

PONS

Appearance......Clear and bright

Colour......Deep gold

Aroma......Sweet and floral, warm banana skin

Taste......Very smooth and silky on the palate, peppery kick on the finish but very little taste

Uses......Salad dressings

Price range......

COMMENTS
Estate bottled Catalan oil.Very low acidity, only 0.25%, it has an elegant texture but how I wish it had more character and flavour.
(FOC)

UNIO

Appearance
Clear and bright

Colour
Golden yellow

Aroma
Lovely immediate aroma of crushed green leaves

Taste
Smooth, light, deliciously flavoursome with just the right touch of pepper on the finish, and deep fruity notes

Uses
Pour it over absolutely everything, it's wonderful

Price range

COMMENTS
An outstanding PDO oil from Siurana in northern Spain. A very complex, sophisticated oil of the highest quality made from Arbequina olives. It's not often you come across oils you could just drink by the glassful, this is one. I have tasted early season and later season oil; fascinating to note the change, but equally delicious. If I had to choose one oil from Spain this would be it. I am hoping a supermarket may be stocking it in the near future. It is staggeringly good value.
(U)

SPAIN

EVO

Appearance......Clear and bright

Colour......Golden yellow

Aroma......Lovely intense leafy with some banana

Taste......Soft and smooth, with the tiniest tickle of pepper on the finish but hardly any taste to describe

Uses......Salad dressings, vegetables

Price range......🌱 🌱 🌱

COMMENTS
Bit disappointing really, not enough flavour or character.

OLIO SANTO

Appearance......Clear and bright

Colour......Yellow gold

Aroma......Fresh and grassy with green apple

Taste......Very delicate, lightly fruity with a tickle of pepper on the finish

Uses......Salad dressings, vegetables, cooking

Price range......🌱 🌱 🌱

COMMENTS
From California's Napa Valley, mixture of Manzanilla and Mission olives. It's well made and quite pleasant but I can't help feeling it has too little flavour to justify buying it, it is a bit on the bland side. There are oils I'd rather have at this price. (OM)

PRATO LUNGO

Appearance
Hazy – unfiltered

Colour
Golden

Aroma
Intense green leaf, salad greens and
rocket with a background note of nuts

Taste
This is such an amazing oil it's like
consuming liquid cream. It has a
sweetish start which moves on to
bitter greens and artichoke

Uses
Salad leaves, salads, vegetables,
dipping oil, grilled fish and meat

Price range
Ⴝ Ⴝ Ⴝ

COMMENTS
This oil proves that when the Americans
put their mind to something they really
master the art. This is oil any Italian would
be proud to produce. It is very Tuscan in
style but with a softer, creamier, easier
approach albeit the olive variety is of
French origin. Outrageously expensive but
what the hell, it's great to be extravagant
once in a while with something as good
as this. It can be ordered via their web site
www.longmeadowranch.com.

USA

STUTZ

Appearance......Hazy – unfiltered

Colour......Gold

Aroma......Bitter leaves and warm garlic. Lovely aroma

Taste......Medium weight, not a lot of flavour or character. Very buttery with a bitter finish

Uses......Pasta and vegetables

Price range......

COMMENTS

Obviously a stylish oil but sadly after the fabulous aroma there was very little character. It is overly buttery and it needs a more bitter cutting edge to be really well balanced and harmonious.

V G BUCK

Appearance......Clear and bright

Colour......Golden

Aroma......Extraordinary and delicious aroma of warm garlic and bitter greens

Taste......Medium weight, smooth bitter taste and buttery finish

Uses......Vegetables, salads, pasta

Price range......

COMMENTS

This oil from Northern California is made from a blend of early harvest green and late harvest black olives. It is well made but it's just a bit too buttery for my taste. I'd like more of the bitter edge that is lurking there to come out. Maybe needs a higher proportion of green olives.

ST MICHAEL SPANISH

Appearance......Clear and bright

Colour......Golden yellow

Aroma......Pleasant astringent tang of green apple and green leaf

Taste......Light flavoured bitter sweet oil

Uses......Salad dressings and cooking

Price range......

COMMENTS
Not a bad example of Spanish oil. Judging by the taste I would guess it's a blend of north and south oil. It's unusually green and bitter for Spanish oil but is good value for money. Tastes sweet and then finishes rather attractively with green grass and a tingle of pepper. Recommended as a good supermarket buy.

ST MICHAEL TUSCAN

BEST  IL

Appearance......Clear and bright

Colour......Golden

Aroma......Pleasantly astringent, crushed green leaves and cut apple

Taste......Medium weight, very nice taste of green apple skin, good finish

Uses......Cooking and salad dressings

Price range......

COMMENTS
A good supermarket oil. Very tasty and one I would certainly recommend. There is also a Tuscan Organic, which is soft and sweet with a taste of grated apple. It is excellent value organic.

ASDA ORGANIC

Appearance
Clear and bright

Colour
Yellow gold

Aroma
Fresh and grassy with a warm fruity
aroma in the background

Taste
Creamy, light, feels good in the mouth
with a pleasant tickle of pepper. Taste of
sweet almonds

Uses
Cooking, salad dressings, grilled fish,
soups, vegetables

Price range
ᵞ

COMMENTS
I have to hand it to Asda with the quality of
this oil at a truly amazing price.
I opened and approached this with some
trepidation because it's so cheap,
I imagined it had to be bad. Well it isn't and
it's very good indeed. I don't know how
they've done it but they are to be
congratulated. If you want to use organic oil
everyday you can at this price.
Recommended.

SUPERMARKET

ASDA EV

Appearance......Clear and bright

Colour......Golden

Aroma......Ripely fruity and sweet

Taste......Rather unpleasant and cheesy with a tickle of pepper

Uses......I wouldn't pour this on my food

Price range......

COMMENTS
Asda manage to find an excellent organic oil and then they serve up *this*. It couldn't be more different.
A really unpleasant oil.

SAFEWAY CRETE

Appearance......Hazy – unfiltered

Colour......Golden

Aroma......Rich fruity aroma of grated apple, quite perfumed

Taste......Creamy in the mouth, lightly fruity with a tickle of pepper on the finish

Uses......Cooking, salad dressings, grilled meat

Price range......

COMMENTS
This is supposed to be single estate oil made from Koroneiki olives, not much flavour or character, certainly not typically Greek in style. It is light, not unpleasant and would be a useful, all purpose, everyday oil.

SAFEWAY GREEK

Appearance......Clear and bright

Colour......Green gold

Aroma......Next to nothing other than a tickle of pepper

Taste......Greasy on the palate, no taste

Uses......I wouldn't

Price range......

COMMENTS
Another example of a horrible tasteless supermarket offering. Where do they find this stuff?

SAFEWAY ITALIAN

Appearance......Clear and bright

Colour......Yellow

Aroma......Faintly fruity

Taste......Light flavour but a powerful pepper kick

Uses......Cooking

Price range......

COMMENTS
Far too much pepper on the finish, otherwise not too bad but there are much better oils around at this price.

SAFEWAY FROM SICILY

Appearance......Hazy

Colour......Yellow

Aroma......Flat aroma of old, over ripe olives

Taste......Flat, tasteless, unpleasant

Uses......I wouldn't

Price range......

COMMENTS

Where did Safeway find this oil? Until I tasted this sample I had
never tasted a Sicilian oil that was anything less than excellent
and delicious. Trust a supermarket to source an example of a
bad oil from a country where it's virtually impossible to find bad
oil. Some poor deluded fool who compiled the label copy
describes the oil as having 'a fresh fruit flavour reminiscent of
green apples and bananas'. Avoid this at all costs.

SAFEWAY FROM TUSCANY

Appearance......Hazy

Colour......Yellowish green

Aroma......Flat, slightly soapy smell of old over ripe olives

Taste......Rather greasy in the mouth, leaves an
unpleasant flavour of cheesy oil

Uses......I wouldn't

Price range......

COMMENTS

This claims to be single estate oil from Tuscany. It is described on
the bottle as 'superior' estate, unfiltered EV made from a blend
including Frantoio. Well, I think it's pretty inferior and I certainly
don't recommend it. It may well be estate oil (though I do wonder
how on earth a single estate could supply the needs of a
supermarket chain) and I should think the estate owner was well
pleased to be rid of it.

SAINSBURY'S EXTRA VIRGIN

Appearance......Clear and bright

Colour......Golden

Aroma......Very faint olive

Taste......Greasy, cheesy, earthy and rustic

Uses......I wouldn't

Price range......

COMMENTS
Not at all a pleasant mouthful. Yet another below par
supermarket offering, from over ripe olives.

SAINSBURY'S GREEK

Appearance......Clear and bright

Colour......Green gold

Aroma......Flat and slightly cheesy

Taste......No taste at all. I could have been drinking
a glass of water apart from the greasy
sensation on the palate.

Uses......Not worth using

Price range......

COMMENTS
The label describes this confusingly as both 'rich and fruity' and
'mild'. It can't be both. Well take it from me it's mild all right. This
is so untypical of a Greek oil, they are usually tasty and grassy,
and yet the label would have us believe it's characteristic of a
Coroneiki (they can't even spell it correctly) olive. Believe me it's
not.

SAINSBURY'S ITALIAN EXTRA VIRGIN

Appearance......Clear and bright

Colour......Yellow

Aroma......Ripe olives

Taste......Really rather tasteless with a bit of pepper on the finish

Uses......Cooking

Price range......

COMMENTS
It leaves the sensation that you've had oil in your mouth but that's the only evidence since the flavour is so negligible. Not worth buying.

SUPERMARKET

SAINSBURY'S SPECIAL SELECTION AFFIORATO

Appearance......Clear and bright

Colour......Green

Aroma......Faint aroma of grass

Taste......Like swigging a glass of water apart from the pepper finish

Uses......Cooking

Price range......

This purports to be stone crushed free run oil from Lake Trasimeno in Umbria. Free run oil is the oil that runs from the olives when they are crushed and prior to pressing, and it's usually delicious. Trust a supermarket to find one without any flavour at all. Not worth buying.

SAN FELICIANO

Appearance
Clear and bright

Colour
Green

Aroma
Very faint hint of green leaf

Taste
Light, gentle silky with just a hint of pepper. A delicate fruity taste which leaves a pleasant lingering flavour

Uses
Cooking, salad dressings, vegetables, fish

Price range

COMMENTS
This oil is packaged for Sainsbury's Special Selection. Attractive light style oil from Perugia. I like it. It is exceptionally good value for money and my tasting sample has found its way into the kitchen for cooking.
(BDV)

FIOR D'OLIVE

Appearance......Hazy – unfiltered

Colour......Green gold

Aroma......Very aromatic and perfumed, crushed leaves and citrus

Taste......Bitter clean flavour of green leaves, with just a little pepper on the finish

Uses......Salad dressings, grilled fish, vegetables

Price range...... 𝄞 𝄞

COMMENTS
Exclusive to Sainsbury's Special Selection.Oil from Garda made from Coratina, Frantoio and Oriarola olives. Delicate oil lacking much character sadly. I prefer the packaging to the oil.
(BDV)

SANTA SABINA TUSCAN

Appearance
Clear and bright

Colour
Yellow

Aroma
Old crushed olives, the kind of smell you get in a mill after a lot of pressing

Taste
Very little taste and what there is, is of slightly old, over ripe olives

Uses
I wouldn't recommend it for anything

Price range

COMMENTS
The label describes this oil as 'strong and full bodied with an almondy after taste'. This is pure wishful thinking. Why do supermarkets do this? It says on the back label '...genuine Tuscany (sic) extra virgin olive oil is only available in limited quantities as the locals usually keep it for themselves..' Easy to see why they decided to part with this one to Sainsbury's. Certainly not worth buying. There are others in the Santa Sabina range which I haven't tasted but if they are anything like this one they are best left on the shelf.

SOMERFIELD

Appearance......Clear and bright

Colour......Yellow

Aroma......Peppery, with bitter green leaf

Taste......Very mild flavour, peppery finish, quite soft and gentle in the mouth

Uses......General cooking

Price range......

COMMENTS
Not a bad example of a bottom of the range supermarket oil.

TESCO EV

Appearance......Clear and bright

Colour......Gold

Aroma......Ripe fruit and green leaf

Taste......Mild sweet fruit with quite a tickle of pepper

Uses......Cooking and robust dishes

Price range......

COMMENTS
It has taste and character and is not a bad everyday oil for general cooking purposes. I would surmise from the flavour that it is predominantly Spanish oil.

TESCO FINEST GREEK

Appearance......Clear and bright

Colour......Green gold

Aroma......None

Taste......less

Uses......I wouldn't

Price range......

COMMENTS
Described on the label as grassy, aromatic with a lingering pepper finish. Do the people who write these descriptions ever actually taste these oils?

TESCO FINEST ITALIAN

Appearance......Clear and bright

Colour......Golden yellow

Aroma......Faint peppery tickle

Taste......Flat and tasteless and rather greasy on the palate

Uses......I wouldn't

Price range......

COMMENTS
Not a good oil at all.

TESCO FINEST SPANISH

Appearance......Clear and bright

Colour......Yellow gold

Aroma......Sweet and fruity

Taste......Not much taste to talk of with quite a peppery burn on the finish

Uses......Not willingly

Price range......

COMMENTS
Suspiciously similar to the Waitrose Spanish. I couldn't decide if they are from the same source because they taste slightly different, the Tesco oil could just be older! Not a pleasant oil. It would be tempting to make a cheap comment about the name of the oil being 'Finest'.

WAITROSE EV

Appearance......Clear and bright

Colour......Yellow gold

Aroma......Rustic, vegetal, over ripe olives

Taste......Greasy, flabby and tired with just the sensation of pepper

Uses......No thank you

Price range......

COMMENTS
Characterless. A bad example of an olive oil. Unusual for Waitrose to be offering oil like this, they are usually so far ahead of the pack in the quality of the produce they offer. None of their own brand oils are presently up to scratch.

SUPERMARKET

WAITROSE GREEK

Appearance......Clear and bright

Colour......Green gold

Aroma......Nothing, faintly cheesy

Taste......Nothing but rather cheesy finish

Uses......I wouldn't

Price range...... ⋎

COMMENTS
Where on earth did they find this? An insult to Greek olive oil.

WAITROSE ITALIAN

Appearance......Clear and bright

Colour......Golden yellow

Aroma......Faint tang of green leaves

Taste......Absolutely no flavour but a burning
pepper finish

Uses......Not willingly

Price range...... ⋎

COMMENTS
Described on the label as a distinctive powerful oil.
Would that it were. Not at all a good oil. I wouldn't buy it.

WAITROSE SPANISH

Appearance
Clear and bright

Colour
Yellow gold

Aroma
Very sweet and fruity, almost like
gum drops

Taste
Very light in style, sweet and fruity,
quite a peppery finish

Uses
Cooking

Price range

COMMENTS
Not entirely enjoyable, it's got too much burn on
the finish to be really good.

The big change since the last edition of this guide is that you can now buy via the Internet. It didn't even figure as an option back in 1995. There are quite a few good food shop sites and I am indebted to Jenni Muir's *thegoodwebguide Food* for introducing me to many I was unaware of.

The other option, which is now much more a part of our everyday lives, is shopping by mail order. I am fortunate that I live five minutes walk from one of the best food shopping streets in southwest London, full of individual shops where I am known and I enjoy the personal contact but in addition I have to confess to loving mail order. It allows you to buy the best of everything even if you don't have a shop nearby. Sitting in the comfort of my home, a glass of wine at hand turning the pages of a well-designed catalogue is my idea of bliss. Food is delivered to your door vac-packed and chilled. In the case of olive oils you don't have to rely on what your local supermarket has to offer you can try a different oil very time you buy if you use mail order or Internet shopping. Every supermarket carries increasingly large ranges of olive oils, often at very low prices but if you want real choice and variety then delicatessens and specialist outlets are the only places to buy your oil. Below are some of the shops I consider to have good selections of oils and where the owners or managers are helpful and knowledgeable. In addition if you know you like Spanish oil or Portuguese oil or Greek oil then the places to go are the specialist ethnic shops, of which thankfully there are still a number around, such as Garcia R & Sons, Lisboa Delicatessen, Athenian Grocery, Fratelli Camisa, Luigi's Delicatessen, and if you live in the capital the guide for these and others is Jenny Linford's absolutely essential *Food Lover's London* (Metro Publications).

If you own a shop and you want to stock any of the oils I have recommended then the list of distributors whose initials are coded on every page under each oil, will allow you to source them. Happy shopping.

LONDON

Oliviers & Co Ltd

114 Ebury Street, SW1
020 7823 6770
and 26 Piazza, Covent Garden, WC2
London has at last its very own specialist olive oil shops. These are a treasure trove of goodies such as glass and chrome cruets for oil, pottery jars for olives, tapenades, chocolates in the shape of olives as well as a wonderful range of excellent oils from France, Spain, Greece, Israel and Croatia, in fact about 20 oils in all but their real speciality not surprisingly is oils from France. You can taste everything and the oils are packed in lovely stylish tins and bottles making them ideal for gifts. I have tasted about half of the range and they were all without exception excellent. Prices range from £6-£12 for 500ml. They have a mail order service from the Ebury Street branch.

Harvey Nichols

Fifth Floor
Knightsbridge, SW1
020 7235 5000
The foodies' haven and still one of the best ranges of olive oils anywhere in London supervised by the knowledgeable Mark Lewis. He is always on the search for new oils to add to the range to keep it exciting and fresh. Worth seeking him out and asking for recommendations. A number of oils exclusive to the shop. Also now a branch in Leeds.

Selfridges

400 Oxford Street, W1
020 7629 1234
Always an excellent food hall, its range of oils is very extensive and second only to Harvey Nichols. They are even stocking two oils from Australia.

Harrods

Knightsbridge, SW1
020 7730 1234
Not as extensive a range as you might expect from a store with a world famous food hall but they do have some oils you won't find anywhere else, as well as own brand and they are one of the few places now stocking one of my favourite commercially blended EV oils from Italy, Sasso, which is sadly difficult to find these days, I couldn't get a tasting sample for this edition.

Fortnum and Mason

181 Piccadilly, W1
020 7734 8040
Who can resist the elegance of one of London's oldest grocers, with its deep pile carpets and chandeliers? They have an excellent own label oil sourced from estates in Tuscany, by David Harrison, the man who produces his own estate oil, Seggiano, as well as a good range of other oils from the Oil Merchant and Guidetti.

Carluccio's

28 Neal Street, WC2
020 7240 1487
Priscilla Carluccio is in charge and she finds truly exceptional Italian oils; see individual entries for Carluccio's range of oils.

Mortimer and Bennett

33 Turnham Green Terrace, W4
020 8995 4145
Sell oils loose. Have an excellent range including a number they import directly and exclusively. Dan Mortimer who runs the shop is a man who knows and is passionate about his olive oils. You'll be in good hands here.

& Clarke's

122 Kensington Church Street, W8
020 7229 2190
The shop next door to Sally Clarke's restaurant is always worth a visit for stocking up on goodies. They import a number of oils direct from producers, so you'll constantly find new discoveries.

The Bluebird Store

350 Kings Road, SW3
020 7559 1000
Terence Conran's store housed in a refurbished 1923 garage is a gastronome's delight and has the kind of extensive array of olive oils you would expect. Their own label oil is utterly delicious and stylishly packaged.

Le Pont de la Tour Oil and Spice Shop
22 Shad Thames, SE1
020 7403 4030
A good range of oils and other groceries as well as interesting cookbooks.

Vinopolis
Axe and Bottle Court
70 Newcomen Street, SE1
London's newest tourist attraction dedicated to all things vinous, not surprisingly carries a range of olive oils too. Well worth a visit.

Hamish Johnston
48 Northcote Road, SW11
My local shop,where I go to buy my cheese and bacon, so I'm bound to be a fan but they do carry a very good range of oils and sell about four or five loose, which is a plus. Also a branch in Abbeville Road, SW4.

The Conran Shop
Michelin House
81 Fulham Road, SW3
020 7589 7401
Perhaps not the first place you'd think of for picking up a bottle of olive oil but they have such an excellent kitchenware department that their range of oils is a sensible and thoughtful addition. Also have a branch in Marylebone High Street, W1.

Villandry
170 Great Portland Street, W1
020 7631 3131
It used to be a tiny closet of a shop on Marylebone High Street, now relocated to airy premises in an area that would be, without this gem, a food wasteland. It's a deli cum eatery, the food is excellent and so is the produce in the shop much of which is organic. I always end up spending too much when I visit.

Panzer's
13-19 Circus Road
St John's Wood, NW8
020 7722 8162
Peter Vogl, the owner, sent me a letter after the first edition came out and asked me to consider including his shop in the directory should there be another edition. Well, we

authors love getting feed back and his shop rightly deserves to be included for its excellent range of oils. A man after my own heart when it comes to careful selection of ingredients.

Mise en place
21 Battersea Rise, SW11
Another local haunt of mine, mainly because they carry the range of Fresh Olive Company of Provence olives and their oils (which I recommend highly), as well as an excellent collection of other oils.

Planet Organic
42 Westbourne Grove, W2
020 7229 1063
As you would expect a large selection of organic oils, many of which I haven't had a chance to taste but their own brand is sourced by David Harrison, who also supplies Bluebird and Fortnum's oils.

Borough Food Market
SE1
Takes place the third Saturday of every month, in and around Borough's old wholesale market. You will find a number of regular olive oil stalls, Brindisa, Waterloo Wine and Guidetti Fine Foods, who usually promote four different oils each month. All available for tasting.

Vivian's
2 Worple Way, Richmond
020 8940 3600
Sadly now no longer with the eponymous Vivian at the helm but still a range of oils to be reckoned with, sold loose as well as in bottle. A shop that locals thank heaven for.

SUFFOLK

Adnam's Kitchen Store
Victoria Street
Southwold
01502 727 222

NORFOLK

Humble Pie
The Green
Burnham Market
01328 738581

NORTHUMBERLAND

The Corbridge Larder
Hill Street
Corbridge
01434 632948

HEREFORDSHIRE

The New Cook's Emporium
21 High Street
Ledbury
01531 632976

WILTSHIRE

Mackintosh of Marlborough
42a The High Street
Marlborough
01672 514069

GLOUCESTERSHIRE

Olives
14 Montpellier Arcade
Cheltenham
01242 245268

CUMBRIA

J&J Graham Ltd
Market Square
Penrith
01768 862281

NORTH YORKSHIRE

Archimboldo's Deli
146 Kings Road
Harrogate
01423 508760

MANCHESTER

Kendals
Deansgate
0161 832 3414

Atlas Delicatessen
376 Deansgate
0161 834 2266

SCOTLAND

Valvona & Crolla
19 Elm Row
Edinburgh
0131 556 6066
I'd move to Edinburgh tomorrow just to have
this as my local shop. People think it's great
being in London for food shopping, well
there's nothing like this in the metropolis.
Amazing range of oils, many exclusive to the
shop and usually at least four or five out for
tasting. Go during the Festival and you may
get to hear Philip Contini performing his
repertoire of Neapolitan songs. You won't get
that in your local supermarket.

Clive Ramsey Delicatessen
28 Henderson Street
Bridge of Allan
Stirling
01786 833903

IRELAND

Roy Fox
49a Main Street
Donnybrook
Dublin 4
01 269 2892

Magills
14 Clarendon Street
Dublin 2
01 671 3830

The Big Cheese Company
Trinity Street
Dublin 2
01 671 1399

Ballymaloe Shop
Shanagarry
Co Cork
021 652032

The Real Olive Company
The English Market Centre
Cork City
021 270 842

McCambridge's
38/39 Shop Street
Galway City
091 562259

Cargoes
613 Lisburn Road
Belfast
01232 665451

Roberts Bakery and Deli
Mold
01352 753119

Blas ar Fwyd
Llanrwst
01492 640215

Vin Sullivan's
Abergavenny
01873 856989

MAIL ORDER

The Oil Merchant
020 8740 1335
fax 020 8740 1319
email: The_Oil_Merchant@compuserve.com
Charles Carey, The Oil Merchant has 27 EV oils on his list. He started importing oils and distributing them to shops long before everyone else. His oils grace any self-respecting shop's range. So if you are a shop or just a customer (since you can order direct) he should be your first port of call.

Clark Trading Company
Order freephone 0800 731 6430
Order freefax 0800 316 0096
www.clarktrading.co.uk
email: cpc@clarktrading.co.uk
Do get hold of their catalogue it is full of gorgeous goodies; pasta, olives, nuts, and charcuterie as well as Nuñez De Prado and Santa Tea olive oils.

Camisa Direct
020 8207 5919
www.camisa.co.uk
This is the mail order arm of Fratelli Camisa, 53 Charlotte Street, London W1, the famous Italian deli. The mail order catalogue has the kind of quality Italian produce you would expect, balsamic vinegars, pasta and preserves and the Santagata and Raineri olive oils. These are both Ligurian oils and generally of good quality as everyday oils. Also offering Oleificio Barbera, a Sicilian oil.

Cucina Direct
0870 727 4300
The mail order arm of Divertimenti

Elanthy
See page 40.
Available mail order on freephone 0800 169 6252. A 3 Litre can is £15.95 including P&P.

WEBSITES

www.deanandeluca.com
Site for the famous US deli, where among other things you can buy Californian olive oils.

www.esperya.com
If you love good food and the best ingredients this is the site for you. It offers more olive oils than any other, and what stunning oils they are. There are 14 exclusive oils. I have tasted four, they were brilliant and the U Trappitu, Sicilian oil is one of the best oils I have ever tasted.

www.fortnumandmason.co.uk
Tuscan and Sicilian EV available.co.uk

www.gourmet2000.co.uk
Sell Carbonell, Iliada, Vallée des Baux oils. Bit of a mess of a site, wrong pictures and no bottle sizes next to some entries.

www.lobster.co.uk
Offer Alziari, River Café, Ravida, Valgiano and Nuñez.

www.longmeadow.com
Home of Prato Lungo oil from Napa.

www.morel.co.uk
Twelve different olive oils, with good notes.

www.olivesetal.co.uk
Sell their olive oils, olives and other delicious foods.

www.oliveharvest.com
EV from Lebanon sold on this site, $24 per litre including p&p.

www.olivesnewzealand.co.nz
Site of The Albany Press run by Jack Hobbs. His olive oil is delicious, tastes of ripe avocado pear. Order some if you want to impress your friends with your far-flung oil acquisitions.

www.theolivepress
Oils from the US, VG Buck, BR Cohn, Lila Jaeger and Spectrum. Gifts, soaps and olive trees for $16.

www.portuguesefoods.com
Oils from Conservas Rainha Santa, delicious Portuguese oils as well as the famous Elvas plums and other regional foods.

www.valvonacrolla.co.uk
The website for Valvona & Crolla, they offer 13 oils on the site.
See entry for the shop in Edinburgh

DISTRIBUTORS

(A) Alivini 020 8880 2525, suppliers of Lungarotti oils as well as other Italian oils
(AO) Anglia Oils 01482 701271 – exclusive agent for Borges oils
(B) Brindisa 020 8772 1600, email:sales@brindisa.com
(BDV) Bocca Della Verita 020 7819 6309, email bdv@parmigiani.co.uk
(C) Carbonell 020 8891 5015
(CAM) Camisa see mail order
(CAR) Carluccio's 020 7580 3050, also see shop entry
(D) Danmar 0800 137064 – distributors of Roi oil
(E) Ehrmanns 020 7359 7466

(EN) Enolio (+39)338.2050286, email: info@enolio.com – importers of 12 Italian regional oils, excellent, informative catalogue
(EW) Eurowines 020 8994 7658
(FOC) Fresh Olive Company of Provence 020 8838 1912, email: sales@fresholive.com
(FW) F.Wilson Ltd 07785 286744 – distributor of Mani oil
(GFF) Guidetti Fine Foods 020 8460 3727 www.guidetti.co.uk – good range of oils including some of my favourites
(LW) Liberty Wines 020 7720 5350, email: info@libertywine.co.uk – the River Café range
(ODY) Odysea 020 7256 8668
(OEA) Olives Et Al 01747 861446, email: sales@olivesetal.co.uk
(OG) The Olive Grove 01275 332423, email: olivegrove@oliveoil.co.uk
(OM) The Oil Merchant 020 8740 1335, email:The_Oil_Merchant@compuserve.com – biggest range of oils available
(P) Peregrine Trading 020 7272 5588, www.seggiano.co.uk – the source of Seggiano olive oil as well as Planet Organic, Fortnum and Mason, Bluebird own brands
(RR) Raymond Reynolds 01663 742230, email: info@raymondreynolds.co.uk
(U) Unio Spanish Arbequina oil is available direct from the producers fax 00 34 977 842298, to William Devin export manager or email wdevin@yahoo.com. The oil presently lacks a distributor here which is a shame.
(V&C) Valvona & Crolla see shop entry
(WW) Waterloo Wine Co 020 7403 7967, email: sales@waterloowine.co.uk

INDEX